Europe's population

Europe's population
Towards the next century

Edited by
Ray Hall
Queen Mary and Westfield College,
University of London

Paul White
University of Sheffield

UCL
PRESS

First published in 1995 by UCL Press

UCL Press Limited
University College London
Gower Street
London WC1E 6BT

The name of University College London (UCL) is a registered
trade mark used by UCL Press with the consent of the owner.

ISBNs: 1-85728-178-0 HB
        1-85728-179-9 PB

British Library Cataloguing-in-Publication Data
A CIP catalogue record for this book is available from the British Library.

Library of Congress Cataloging-in-Publication Data are available.

Front cover: Ndoj Zeff Nika, 18 months old (Rhodri Jones/Panos Pictures).

Typeset in Bembo.
Printed and bound by
Biddles Ltd, Guildford and King's Lynn, England.

Dedicated by the editors to their own children:

Robert and Rosamund Hall
Katherine and Helen White

Contents

Preface

In the final decade of the twentieth century, two common themes of public debate and of academic discussion in the social sciences have concerned futures research, and the European scene in the context both of developments in the European Union and of post–Cold War changes in other parts of the continent. At the 1992 Annual Conference of the Institute of British Geographers, the Population Geography Study Group organized a session on the future of population change in Europe, bringing together these two major themes in the context of demographic change. The conference session elicited a considerable amount of interest, and it was there that the initial idea for this current volume was first discussed.

The resulting book is actually somewhat different in structure and content from the original conference session. Certain contributions to the conference have not carried forward to the book, whereas new authors have been commissioned to produce chapters on topics not originally covered. Where certain chapters originated in conference papers, they have been rewritten for the purposes of publication. The chapter by Maria-Carmen Faus-Pujol is based on the paper she originally gave as the first John Coward Memorial Lecture, in honour of the distinguished scholar of population geography who was killed in the Kegworth plane crash of January 1989.

The aim of *Europe's population: towards the next century* is to contribute to informed discussion of the demographic futures of Europe as a whole. The whole range of population geography is covered, including considerations of fertility and mortality, household and family structures, labour-force issues, population redistribution and international migration. The authors were each asked to look to the year 2000 and, where possible, beyond. The approach adopted eschews highly technical projections, instead highlighting issues and alternative scenarios within general contexts of societal and economic evolution. The authors have been drawn from several European countries, and the intended coverage is Europe-wide, although in certain chapters the paucity of current data from some countries (especially in eastern Europe) narrows the discussion to the countries of the European Union.

The editors express their thanks to the contributors for their work, and to the technical staff of their respective institutions for cartographic and typing support.

October 1994

xi

Contributors

Tony Champion, Department of Geography, University of Newcastle, Newcastle upon Tyne, UK.

Maria-Carmen Faus-Pujol, Department of Geography and Spatial Planning, University of Zaragoza, Ciudad Universitaria, Zaragoza, Spain.

Allan Findlay, Department of Geography, University of Dundee, Dundee, UK.

Anne Green, Institute for Employment Research, University of Warwick, Coventry, UK.

Ray Hall, Department of Geography, Queen Mary and Westfield College, University of London, Mile End Road, London, UK.

Markku Löytönen, Department of Geography, University of Turku, Turku, Finland.

Maura Misiti, Institute for Population Research, Viale Beethoven 56, Rome, Italy.

Calogero Muscara, Department of Spatial Planning and Urbanism, University of Rome "La Sapienza", Via Cassia 32, Rome, Italy.

Daniel Noin, Institute of Geography, University of Paris I (Panthéon-Sorbonne), 191 rue Saint-Jacques, Paris, France.

David Owen, Centre for Research in Ethnic Relations, University of Warwick, Coventry, UK.

Pablo Pumares, Research Centre for Economics, Society and the Environment, Calle Pinar 25, Madrid, Spain.

Vicente Rodriguez, Research Centre for Economics, Society and the Environment, Calle Pinar 25, Madrid, Spain.

Deborah Sporton, Department of Geography, University of Sheffield, Sheffield, UK.

Paul White, Department of Geography, University of Sheffield, Sheffield, UK.

CHAPTER 1

Population change on the eve of the twenty-first century

RAY HALL & PAUL WHITE

The context for demographic change

Population events have significance at a wide diversity of scales. Childbirth, marriage, divorce, illness, death and migration are all intensely personal occurrences which are imprinted in the biography of those concerned – indeed, in its simplest sense, a human biography consists simply of demographic events: "hatches" and "despatches" (with "matches" on the way in many cases). Demographic events are central to the great turning-points in the life course, and have been recognized by anthropologists as being crucially ritualized as "passages" from one life-condition to another. European societies have generally ceased to celebrate the onset of puberty, menarche and menopause (these being specifically biological developments and thus permissive, rather than explicitly demographic), but most other demographic events are marked in societal ways – christenings and naming ceremonies, weddings, house-warming parties, funerals. And certain ethnic groups within present-day Europe have further societal acknowledgements of, for example, menarche in girls (Islamic societies) or arrival at adulthood (the Jewish bar mitzvah for boys, bat mitzvah for girls). Demographic events are thus overladen with societal and cultural meanings in many highly significant ways.

At the same time, many of these events, taken in aggregate, are of crucial significance in the evolution of wider structures of life. Births and in-migrations increase population sizes, whereas deaths and out-migrations reduce them. Given that such events are not generally randomly distributed throughout whole populations, the outcomes of population processes serve to produce a situation in which the components of populations (by gender, age, social status, ethnicity, and so on) are in a constant state of flux. Not only do demographic events taken in aggregate change population totals, they also change the size of the subgroups that make up such populations. Fertility, nuptiality and the propensity to migrate all show differentials

1

between, for example, social classes and between different ethnicities. Death, apparently the great leveller, actually spares certain people longer than others.

These variations in the occurrence of birth, marriage, migration and death are themselves influenced by a complex series of contextual forces. Single events within individual families take place against a background of economic, social, cultural, educational and political, as well as personal, circumstances. Apparent elements of choice are, in reality, often highly constrained. In the demographic sphere human beings cannot be completely existential: they have neither the ability nor, perhaps, the desire to construct their own code of demographic behaviour. Societal norms and fashions are important, and replication of role models and "expected" behaviours are common.

To analyze population change at the aggregate level requires consideration of these broader contextual circumstances. Without it we can deal only with descriptions. Demographers have a range of highly sophisticated, largely statistical, models for investigating population change, yet the success of these models has been limited by a general reluctance to relate them to the evolving economic, political and societal contexts within which real populations operate. The real world does not come as "clean" as many mathematical models would like. This does not mean that demographic analysis is a fruitless task: it does mean that demographic predictions must be supplemented by more qualitative and evaluative considerations of circumstantial developments. Nowhere is this more important than when we look to future changes in population characteristics.

The significant issues in fin-de-siècle Europe

Throughout history there has been a tendency to regard the last decade of a century as in some way heralding both the end of the old and the instigation of a new beginning in a particularly significant fashion. Although we may not all hold a millenarian view about the coming of the year 2000, there seems nevertheless to be some special importance in looking forward into a new century. Within Europe there seem to be good grounds for doing so, related to some major developments and events of the latter years of the twentieth century. We should briefly review these. However, what is possible in an introductory chapter to an edited collection such as this is only a superficial listing of the major influences. Europe is a complex continent displaying a series of tensions between tendencies towards unity and leanings towards fragmentation. These influences are by no means present at all places, and may work differently in different societies and political systems.

POLITICAL CIRCUMSTANCES

The series of rapid political changes that resulted in the demise of the communist regimes of eastern Europe between 1989 and 1991 gave rise to a degree of triumphalism in certain Western circles. One of the most controversial (and misunderstood) statements arose through Francis Fukuyama's suggestion that this was "the end of history" (Fukuyama 1989, 1992). This concept was intended to apply to the triumph of liberal democracy and the overall hegemony of capitalist free-market economic structures. In this Whiggish view of history, human endeavour is argued to have evolved to its conclusion in terms of the acceptance of a uniform set of underlying structural forces as the determinants of global political economy. Leaving on one side the possibility that this is a narrow view that does not apply to large swathes of the less developed world, at least in political terms, the question has to be asked as to whether the end of "history" is now with us, and what this would imply. In global geopolitical terms post-war history has been written through superpower conflict which is now over, to be replaced by the concept of a "New World Order".

Superpower conflict lay at the heart of political change and development in post-war Europe. A strong case can be made out for the post-war division of Europe as being behind such important developments as the Marshall Aid plan (launched by the USA to rebuild the democratic capitalist economies of Europe as bulwarks against communism); the creation of NATO; and the movement towards western European unity from the European Coal and Steel Community, through the European Economic Community to the European Union. Without the perception of an external political threat it is possible that the pace of the drive towards European unity will slow down. It is perhaps ironic that the European Community embarked on the ambitious task of deepening the relationships between the member states just at the time when political and consequent economic change in eastern Europe ushered in the competing issue of widening the Community (Singer & Wildavsky 1993). Despite the Maastricht Treaty, the years since 1989 have generally seen a weakening of the arguments for deepening the EU, and Germany in particular has reoriented much of its economic gaze away from the benefits of economic and political integration westwards and towards both the internal problems of reunification in 1990 and the possible economic pay-off from enlarging the Union and from investment in the former eastern Europe.

Even if Fukuyama's major thesis is correct (and Fukuyama is not arguing that history with a small "h" will come to an end), this does not mean that geography will become an irrelevance. Indeed, as several have argued, states and territories may be redrawn and new political movements develop (Johnston 1994, Pattie 1994). The hegemony of liberal democracy and capitalism does not imply the end of geography. Capitalist development is spatially uneven development, and the resultant

3

economic imbalances between economic areas lie at the root of some spatial flows, of which population migration is one of the most important.

A further outcome of the events of 1989–92 has been the removal of centralizing state forces in much of eastern Europe and the re-emergence of regionalism and ethnicity as crucial factors in political evolution. We have seen the break-up of the former Soviet Union, the splitting apart of Czechoslovakia, and the conflicts within former Yugoslavia. Each of these episodes has held, or will hold in the future, the potential for demographic consequences, particularly through migration, sometimes of refugee movement.

These are large-scale, often international, issues. Within individual states both national policies and emergent political ideologies have important effects. Demographic changes are influenced both directly and indirectly by government policies. The state plays crucial roles, through education, health, employment and social security policies, in influencing patterns of fertility, nuptiality and mortality. The exact nature of the effects are often poorly understood, reflecting the complexity of the interface between aggregate influences and individual activities that create demographic events. Throughout Europe governments of the late twentieth century have become aware of the implications of population ageing and possible declines for labour supply (Anon 1993) or for government expenditure, for example through pensions (Bichot 1993).

As important as actual policies, however, are trends in ideology within contemporary Europe. These are strongest in western Europe, but with the increasing Westernization of the East they are likely to spread. Common throughout are features such as a rising emphasis on individualism, the privatization of formerly community interests and ownership, and the spread of consumerism as the basis of individual ambition. Although many of these trends emanate from New Right political developments of the 1980s, certain political commentators of the right have begun to repudiate the outcomes and to call for a revival of the "common life" (Gray 1993). An increasing emphasis on "self" could be argued to lie at the heart of declining fertility trends over much of contemporary Europe, and any general vision of a dramatic return to above-replacement fertility seems unlikely, despite recent developments in Sweden where this has, in fact, come about.

In a very interesting exercise a group of social science commentators were recently asked to build scenarios for the future of Europe in the year 2020 under three major and competing political ideologies (Masser et al. 1992). The first was labelled a "growth scenario" and envisaged the dominant political goal as being economic expansion; the second, labelled an "equity" scenario, in which policies aim at reducing social and spatial inequalities; and a third, "environmental" scenario, associated with quality of life and "green" issues. The expert panel felt that the growth model was likely to dominate political thinking, and that in terms of its impact on population some interesting developments were likely. Population ageing

4

would continue rapidly and continuous increases in productivity would be needed to sustain wealth generation, since the proportion of the population making up the active labour force would be falling. Average household sizes would continue to decline, with further family fragmentation and life-styles dominated by individually based consumerism. In such circumstances labour shortage would be a real concern, and could only be met by large-scale labour in-migration from elsewhere: demographic modelling suggests that the scale of inflow needed would be considerable (Lutz 1991).

The equity scenario also created a need for immigration, although at a lower level, and a reassertion of collective values (associated with possible increases in fertility) was seen as possible. However, it was only the environment scenario that was felt to lead to stability in labour demand and supply, with the possibility of an increase in fertility levels once again to a level where immigration would not be needed.

At every scale, future political change will prove a crucial controlling influence on demographic change in Europe, just as it has done over the recent past. However, political developments will themselves take place within an economic context that, at present, presents a somewhat confused picture.

ECONOMIC CIRCUMSTANCES

Much has been written about contemporary economic globalization; new international divisions of labour; the switch from production to consumption as an economic basis in various western European core economies; and the potential for a fifth Kondratieff wave of growth. Much established analysis of these phenomena in the European context took place under the East/West division of the continent or in the early period of political change (Fröbel et al. 1980, Wallerstein 1991). The European implications of economic restructuring, however, now include the processes of absorption of newly market-orientated economies in east-central Europe into a wider economic system. This is already bringing considerable disruption at several scales, altering the life-chances of individuals and influencing them in demographic decision-making.

Economic restructuring is spatially biased, with patterns of winning and losing sectors and regions. It should not be thought that regional inequalities were unknown under state socialist systems, but as parts of eastern Europe are absorbed into the wider European space-economy as semi-peripheries the overall patterns of regional economic wealth and growth are likely to alter. Germany becomes even more of a central pivot within the European economy, and once the economic problems of German reunification are out of the way there is likely to be a rapid redrawing of the familiar maps of economic potential and location in Europe

5

(Keeble 1989). One further aspect of such change is likely to relate to technological developments, for example in terms of super-fast rail networks, or further innovation in telecommunications and the information economy (Hall 1987). Places are likely to be drawn together in time and ease of communications and access, while at the same time remaining different in economic structure, political orientation and social and ethnic complexity.

THE INDIVIDUAL AND SOCIETY

Reference has already been made to increasing features of individualism and the fragmentation of earlier versions of community and society. Given the demographic significance of these issues, as highlighted at the start of this chapter, key future influences on population change will result from various of the tensions identified.

The movement of many European economies into a post-industrial phase has been accompanied by the rise of post-modernity as a social and cultural phenomenon, and by the validation of certain quasi-existentialist terms of validation for individual and group behaviour. Gender equality is more apparent in some areas of Europe than others, whereas racism is a general feature that is expressed differently in different societies; attitudes to alternative sexualities are variable. Changes in the future labelling of core and peripheral (vulnerable or marginalized) elements in society will have important but unpredictable influences on population change, affecting, for example, future household constructions, fertility, nuptiality, the possible demographic "assimilation" of ethnic minorities, and the committal of resources to certain defined "problems".

Many commentators agree that social polarization will be an increasing issue over the coming decades (Hamnett 1994), resulting from the coalescence of economic, political, ideological and social trends. Certain groups and individuals get an increased level of choices, whereas others face highly constrained futures with little real prospect of improvement in their conditions. The geographical significance of social polarization is manifest in phenomena such as counterurbanization and gentrification, in both of which processes those in society with the economic power to satisfy their preferences for residential locations are able to do so whereas those with no such power become ghettoized into declining industrial areas, suburbs and inner-city districts of economic stagnation, under-investment and decay. Although the future may not hold a continuation of counterurbanization as we have seen it over the past two decades, the basic element of individual decisions on location on which it has rested will still be present for many in society.

With the reduction in state intervention in housing markets, and with general trends of privatization in service provision, the overall levels of support for the poor and for other marginalized groups will decrease, at a time when the size of such groups will almost certainly increase (for example through ageing, and through the growth of ethnic minority communities). Social issues will therefore reflect demographic changes as well as political and economic forces.

THE WIDER CONTEXT

European changes have a global context, and it is important that we should recognize that forces and influences in distant locations will have an impact on European population changes. Recent political change has not been confined to eastern Europe. A feature of the early 1990s has been the move to democratization in Africa (which might in part be used to support Fukuyama's thesis), which may change the course of the economic structures as well as the political realities of that continent. Elsewhere what has been characterized as a rise of Islamic fundamentalism has significance not just for the rimland countries around Europe but for many of the populations of immigrant origin living within (predominantly western) Europe itself.

Two books by the historian Paul Kennedy have explored some of the wider themes that may underpin future global change. In the first of these Kennedy (1988) explored the relationships between economic and military strength since the sixteenth century, and argued that the ultimate costs of the desire to retain military supremacy lie in the diversion of an increasing proportion of resources into defence expenditure and the slowing of economic growth. Writing in the late 1980s, Kennedy saw this process as currently operating in the USA, whereas in the Asian-Pacific realm economic growth was leading to the creation of economic, without military, hegemony for the Japanese: the economic centre of gravity for the world economy might increasingly shift to East Asia, with economic consequences for Europe and the USA.

Since *The rise and fall of the Great Powers* was published, the end of the Cold War has brought a new situation. The "peace dividend", painful at first since it might involve large-scale unemployment in certain areas of the military–industrial complex, could produce a reorientation of productive innovation and renewed economic growth in wealth-generating industries in Europe. Against this, however, must be placed the two phenomena already referred to: the rise of consumption rather than production as the economic base, and the incipient decline and ageing of European populations.

In his more recent book, *Preparing for the twenty-first century* (1993), Kennedy puts demographic issues at the forefront, starting with a re-examination of the relevance (perhaps postponed until now) of Malthus' ideas at the world scale. Kennedy sees western Europe's own demographic problems as being the "greying" of its own population, and the new pressures set up by circumstances operating outside the region – in eastern Europe and beyond. As Kennedy puts it:

> Far from being something that Europe can safely ignore, global demographic trends can affect the social order, delay (or reverse) the opening of the EC's internal barriers, and even influence its foreign policy. (Kennedy 1993: 276)

The extensive publicity given in the Western media to the 1994 World Population Conference in Cairo attests to the growing salience of demographic issues on the political agenda.

Predicting population change

Demographers have available to them a highly sophisticated set of mathematical and statistical models on which to base discussions of future developments. However, the technical detail of such projections often masks the fact that at root they are all based on several assumptions about the continuation of certain aspects of present patterns, or about changes to current demographic rates (such as fertility or mortality) that will occur in predictable and stable directions.

Many projection exercises in the past could be criticized as being demographically sophisticated but naïve in terms of their conceptualization of the economic, political and social context for future developments. Projections of future migration have been attempted relatively rarely in comparison with those dealing with fertility, mortality and age-structure. In recent years the narrow task of demographic projection has been broadened through the adoption of a wider social scientific approach in which possible social, economic and political scenarios are built up and explored in terms of their possible impacts on demographic variables. Examples of the discussion of such scenarios in a non-quantitative fashion have been given earlier in this chapter (Masser et al. 1992). The development of more quantitative scenario building has been associated with a research group working at the International Institute for Applied Systems Analysis (IIASA), and this has resulted in several major studies (Lutz 1991, 1994, Lutz & Prinz 1993).

Scenario building nevertheless can still only retain a limited number of non-demographic variables in its work: the choice of these, and the development of plausible event-histories in, for example, governmental policy-making or economic growth rates is an art rather than a science. Where scenario building is useful is in suggesting the range of possible outcomes without placing undue stress on any one specific likelihood. Scenarios are to be used as a basis for discussion rather than as a basis for planning. Discussions in several chapters in the remainder of this book use a scenario framework in which demographic variables are seen as nested within political, economic and societal contexts which themselves may be variable over time.

However, before turning to consideration of the structure of the rest of this book it is instructive for the social scientist to turn to the more imaginative works of fiction that have gazed into the future of population issues. Several of these raise important questions that, although still apparently within the realm of impossibilities, nevertheless contain dimensions that merit discussion.

Headbirths by the German novelist Günther Grass caused a stir when it appeared in 1980. Grass contrasts the low fertility of Germany with the prolific population of China:

> The Germans are dying out. Living space without people. Is such a thought possible? Is such a thought permissible? What would the world be like without Germans? Would it have to look to the Chinese for salvation? Would the other nations of the earth find they lacked salt? (Grass, English translation 1984: 11)

In this piece of fiction Grass explores both the ironies of depopulation, and also the individual and societal context in which couples decide not to have children (or rather, fail to decide to dispense with contraception used as a norm and instead to seek conception). Grass creates a couple (significantly both are geography teachers), worrying endlessly about the state of the world and their own moral responsibilities in a world of soaring population growth rates:

> She harps on data, curves, computer projections: 'Here, look at South America. Everywhere a three per cent increase. Five in Mexico. What little progress there is, they eat it up. And the Pope, the idiot, still prohibits the pill. She takes it regularly. Always at the start of her first class. A kink, or call it a kinky demonstration of her rationalized self-denial. So this is how the *Headbirths* movie could begin: Long shot of the Indian subcontinent. She, cut off at the waist, covering half the Bay of Bengal, all Calcutta, and Bangladesh, casually takes the pill, slaps a book shut . . . and says "It is safe to say that birth control as a means of family planning has been a failure in India" (Grass, English translation 1984: 14–15).

Grass sees low fertility as brought about more by altruism than by individualism, but his discussion explores many of the personal issues involved in a way that has not been undertaken by social scientists.

Aspects of future fertility have also been explored in two recent French novels. In *2024* by Jean Dutourd (published in 1975) the author deals with a world of the elderly, brought about by the almost total cessation of fertility. Such scenarios are, of course, inherently improbable, but they allow the imaginative discussion of the relationships between demographic structures and the whole workings of society. More recently Amin Maalouf, in *Le premier siècle après Béatrice* (1992), has created a modern parable in which, with the possibility of choosing the gender of a future child, there is a rapid reduction in the birth of girls in the countries of the South (the less developed world). Although governments of the North (the developed world) ban the vaccine that eliminates female births, the gender imbalance in the South brought about by the ancient desire for male children creates worldwide disturbance affecting Europe (much of the novel is set in France) as much as elsewhere. Again a fictional device has been brilliantly used to explore the relationships of simple demographic facts to wider social and cultural issues. Maalouf's demographic scenario is not totally implausible: we are on the verge of parental sex choice of unborn children. Similarly P. D. James (1993) has recently taken an actual biological fact – the rapidly dropping sperm count in males – and extrapolated this to a depopulating world with almost zero fertility. In novels such as these creative writers have explored demographic futures in a particularly provocative manner.

Finally, it is interesting to note that in another *fin de siècle* period, that of the nineteenth century, a futurist demographic novel was written dealing with the prospect of longevity becoming so burdensome on society that compulsory euthanasia is introduced to retain a satisfactory demographic balance: the author of this story of *The fixed period* was Anthony Trollope (1882). This is not a theme that has been taken up in more recent fiction.

The crucial demographic issues

When we examine the complexity of the European scene at the end of the twentieth century with demographic changes in mind several themes emerge. In particular, the slowing of population growth rates in Europe is of great significance not only to that continent but also to the world as a whole. European population change in the nineteenth and early twentieth centuries provided the evidence for the so-called theory of the demographic transition, and it is perhaps in Europe that we can see the final stages of the model being worked through. What happens in Europe may indeed prove to be a blueprint for the rest of the world.

Europe's population growth has slowed to unprecedentedly low levels as a result of great improvements in life expectancy and the decline of fertility to replacement level or below. Projections suggest that western Europe's population will be in absolute decline by 2005 and southern Europe's by 2010: meanwhile the populations of northern and eastern Europe will continue to grow very slowly (United Nations 1991).

Such declines have very profound sociodemographic implications. Europe's share of world population is falling: in 1950 it stood at 15.6 per cent of the total, by 1990 it was 9.4 per cent, and it is projected to fall to 6.1 per cent by 2025 (United Nations 1991). These broad trends are crucial to the future role of Europe in the world of the twenty-first century. The remainder of this volume explores some of the more specific demographic components in detail in the search for views of the population landscape of Europe over the next ten to twenty years.

A NO-GROWTH POPULATION

The key demographic issues facing Europe are highlighted in Chapter 2 where the implication of a no-growth population for changing demographic structures is the key theme. A model of population structure in the mid twenty-first century is developed which is markedly different from that of the late twentieth century. Whether such a population structure actually comes to pass will depend on what actually happens to fertility levels in Europe: the model is based on the assumption that fertility will remain at very low levels, according with many of the arguments made earlier in this chapter – that increasing fertility is unlikely under current societal and political structures. Nevertheless, fertility may increase, one factor which could be important here being the continuation of immigration into Europe from higher fertility regions of the world. Attitudes towards immigration may well change, especially as indigenous populations begin to decline in some regions of Europe so that labour shortages become more pronounced. There is then the question of how immigrant and ethnic minority populations will behave with respect to fertility. Will successive generations adopt the low European fertility norms, or will they maintain higher fertility (for example through their marginalization and non-integration)? If the latter, would this be enough to produce some overall population growth at national levels?

Other questions raised in Chapter 2 concern whether an ageing population need necessarily be seen as a "problem". The population may be older, but it may be technically more competent and healthier. Adjustments in attitudes towards ageing are needed, and indeed we should redefine what we mean by old age as life expectancies increase.

FERTILITY AND HOUSEHOLD CHANGES

Succeeding chapters develop and expand on points raised in Chapter 2. Whether fertility might in fact increase is one of the themes explored in Chapter 3, which considers how changing attitudes towards both sexuality and the role of women

11

contributed to the dramatic declines in fertility from the mid-1960s onwards in most of western and northern Europe. The changing role of women, particularly in employment, combined with women's developing sense of individualism, make it unlikely that they will ever return to having large families. Indeed, very low fertility is likely to remain the norm unless there are dramatic public policy changes in respect of childcare provision and equally dramatic changes in male attitudes towards their own childcare responsibilities.

The changing role of women which has had such a profound impact on fertility has similarly contributed to changing attitudes to relationships – particularly marriage – and living arrangements. So we have seen a rise in divorce rates, a growth in cohabitation, and an increase in births outside marriages since the early 1970s, all of which impinge closely on household structures. The dominance of the nuclear family household of the mid-1960s and earlier has given way to a greater variety of household types: for example, people living on their own now make up about one-quarter of all households in the European Union compared with about 18 per cent in 1970. This in itself is an indicator of that greater individualism discussed earlier in this chapter. In the future households are likely to become yet more varied and flexible in structure and type.

LABOUR MARKET CONSEQUENCES

The labour market implications of an ageing population and changing population distributions within the context of likely changes in labour demand forms the subject of Chapter 4. Likely population structure changes are explored in detail, which shows that there will be no major labour force decline in Europe as a whole before 2025, although regional variations and detailed changes in the age structure of the labour force will be significant. In particular there will be an ageing of the work force which may have an impact on the rate of adoption of new skills. At the same time, spatial changes in the distribution of labour will have implications for the future prosperity of regions as less prosperous areas lose labour to the more economically dynamic. Patterns of labour demand are also likely to change, with a decline in manual occupations and a rise in professional and managerial jobs. Again, regional variations will result in some areas facing short or long-term problems of unemployment while others experience shortages of suitable labour. Coping with these changes will pose a range of problems, and life-long education and training will become more important, particularly for disadvantaged groups and areas.

Broader labour force changes are also highlighted, particularly the emergence of segmentation between core and peripheral workers and the feminization of the labour force. Europe has several labour force options available, and the outlook need not be gloomy: however it is important that the work force, whatever its size or origin (and immigrant origins are further possibilities), should master techno-logical change. A highly skilled labour force rather than a large labour force is going to be an essential ingredient for the twenty-first century.

MORTALITY AND MORBIDITY

Chapter 5 explores the spatial inequalities and trends of mortality in the countries of the European Union and considers the prospects for the year 2000. There are still considerable variations in life expectancies within the EU, which are even more marked if regions rather than countries are considered – some countries show con-siderable internal inequalities. The sexes also show pronounced inequalities in the face of death throughout the EU.

Explaining these patterns is not always easy, but factors such as diet generally seem to be of greater significance than health services. Life expectancies are expected to continue to improve and should reach around 80 years for the popula-tion as a whole throughout Europe in the next century. At the same time spatial disparities are likely to narrow although differences between social groups and between the sexes will be more resistant to change. Will these longer life expect-ancies be achieved by lengthening the period of healthy, active life (which would be a positive step, helping to alleviate some of the fears and problems of the demo-graphic time-bomb discussed in Ch. 4), or will the health-care burdens increase and put ever greater pressure on social welfare budgets and on the tax-paying work force?

Chapter 6 explores a disease unknown before the late twentieth century, AIDS, and its likely impact on overall mortality in Europe. AIDS is a reminder that the future always remains uncertain. By the beginning of the 1980s the belief was growing that medicine would ultimately conquer disease, but the arrival of AIDS causes this belief to be questioned. The scenario developed in Chapter 6 is reason-ably optimistic, arguing that the effect of the HIV epidemic on demographic trends in Europe may be no more than marginal. Nevertheless, the health service costs are likely to be huge, which poses further problems for health planners facing escalating demands for increasingly sophisticated health care, demands which are likely to increase as populations age.

COMPLEX PATTERNS OF MIGRATION

The second half of the book deals with migration: a demographic variable likely to be crucial in determining the shape and structure of Europe's population in the next century. As Chapter 7 emphasizes, migration is the dominant explanation of differential population growth between regions, although patterns of growth and decline are often much less clear-cut than used to be the case in the days of straightforward rural–urban movement. Patterns of internal migration are likely to become even more diverse in the future, although the range of possible outcomes is unclear. As patterns of work and occupational histories change so patterns of migration will reflect the variety of local and regional economies. In order to make any predictions about migration flows and resulting population distributions we need to make assumptions about future scenarios of growth, equity or environmentalism, as discussed earlier in the present chapter.

Chapter 8 develops some of these ideas further in relation to the migration of highly skilled workers, an aspect of migration that will become of increasing importance as employers demand workers with very specific skills. As Chapter 4 makes clear, the real barrier to future European economic development will not be overall labour shortages as such, but specific skills shortages. Thus, as demands and opportunities increase, the highly skilled are likely to be amongst the most mobile in the population. So far their migration is relatively and surprisingly limited. There is still strong cultural resistance to skilled international migration in northwestern Europe. It seems unlikely that any economic integration within the EU will occur at the expense of any major erosion of cultural differences between member states, despite some political rhetoric to the contrary. Nonetheless, skilled migration may well become more important by the start of the next century as the young, ever more familiar with the languages and cultures of other European states, become increasingly willing to circulate internationally. Schemes such as ERASMUS and SOCRATES, encouraging international student mobility, may in turn lead to changes in attitudes to further mobility. Such cultural changes are necessary since the lowering of economic barriers to skill transfers is unlikely to be sufficient to produce a massive increase in professional and managerial migration.

Two particular aspects of current concern in relation to migration are explored in the final two chapters: east–west migration within Europe, and south–north migration to Europe. The first relates to pressures from eastern Europe and particularly the former Soviet Union for migration to western Europe. There was great concern about large numbers of would-be migrants who would wish to move westwards once free movement was possible from the former Soviet empire. Although the exaggerated fears of the early 1990s are now seen as unfounded, there is nonetheless the possibility of a continual seepage of people frustrated by economic problems in the countries of eastern Europe.

The pressures for migration from South to North are unlikely to go away since the rapidly increasing populations of the developing world seek to improve their life-styles, and those European countries fringing the Mediterranean are likely to feel the greatest pressures. These are now countries of immigration, contrasting with their role as labour reservoirs of only 20 years ago. Much of the immigration to countries such as Italy or Spain is undocumented, so that the true scale of contemporary movement is unknown.

Pressures in southern Europe are very much a function of the rapidly increasing populations across the Mediterranean. Southern Europe at present has 144 million people, and North Africa 157 million. By 2025 southern Europe's population is projected to be almost unchanged at 148 million, while that of North Africa will have almost doubled to 280 million. North Africa currently has a total fertility rate (TFR) of 4.5 which, combined with a low death rate, implies a doubling of the population in 28 years. By contrast, in southern Europe the total fertility rate is the lowest of any region in the world at 1.4, significantly below the figure required for replacement of the generations.

Fortress Europe or Fortunate Europe?

The proximity of regions of such extreme demographic contrasts brings into sharp focus the problems that will increasingly face Europe in the twenty-first century. Can, and should, Europe become a fortress against the rest of the world? How permeable will her frontiers be, the efforts of frontier control and immigration officials notwithstanding? New patterns of migration are emerging in southern Europe which relate strongly to the developing post-industrial and post-Fordist trends in European economies as a whole, and to aspects of the new world economic and political order affecting Third World countries. Labour markets are becoming more fragmented and segmented, and immigrants are more widely distributed geographically in receiving countries throughout Europe. The very low birth rates in southern Europe in particular mean that migration may well become more important as a means of coping with labour force problems. But even if migration from outside Europe were totally restricted, movement between European states (members of an enlarged EU) is likely to increase, with the Sunbelt states of southern Europe as beneficiaries.

Europe is not an island or isolated continent without a context. It has innumerable worldwide economic, political, social and cultural connections. To its east and south the demographic pressures are intense. Europe's problems of a declining and ageing population with high material demands and expectations have to be faced in the context of a still rapidly growing world population that looks to Europe and elsewhere in the developed world for its expectations of what can be achieved in material progress. And where Europe has led demographically in the past, other regions have followed: other world regions may again follow Europe's demographic development in the twenty-first century.

Europe is entering uncharted demographic territory, and the following chapters

15

focus attention on some of the key issues which European policy-makers have to grapple with in the coming decades, highlighting current trends and speculating about possible future developments.

CHAPTER 2

Changes in the fertility rate and age structure of the population of Europe

MARIA CARMEN FAUS-PUJOL

Introduction

Europe has been a pioneer of the great demographic changes that have taken place in the world over the past two centuries. Rapid population growth characterized the continent during the late eighteenth and nineteenth centuries while today the European population finds itself in a period of very low or zero growth which some see as the last stage of the demographic transition, and others as the beginning of a new process of demographic change whose consequences are difficult to predict.

Europe's population growth rate was low throughout the interwar period, a situation which caused concern in many countries. However, at the end of the Second World War the birth rate rose throughout Europe and in the immediate aftermath of the war western Europe, and Germany in particular, received some twenty million displaced people. The economic boom of the 1950s and 1960s created enormous demands for labour which were satisfied by an influx of immigrants to western Europe from the countries of the Mediterranean: Spain, Greece, Turkey and North Africa. The labour influx was halted as a result of the oil crisis in the early 1970s. Around the same time it was clear that the decline in the birth rate that had begun in many countries in the mid-1960s was continuing. The late 1960s and early 1970s can therefore be seen as a demographic watershed.

The changes that are currently taking place in the European population are of concern not only to demographers, sociologists, economists and geographers, but also to politicians responsible for the creation of what is now known as the European Union (EU). These changes in population include three aspects which should be highlighted over and above the others, namely the sharp fall in the fertility rate, the stagnation and relative ageing of the population, and the migration of non-Europeans to Europe – a reversal of the traditional flow of intercontinental migration that characterized Europe in the nineteenth and early twentieth centuries.

Although each of these changes merits a detailed study of its own, this chapter concentrates on giving an overall view of the current demographic situation in

Europe with particular reference to the fertility rate and the age structure of the population. It examines in detail four particular aspects of Europe's population: (a) the idea of European space and population; (b) the growth of the population; (c) the evolution of fertility; and (d) the consequences of all the above on the population structure in the next century.

The European space

Geographical facts are inseparable from the space in which they occur. For this reason, before examining European fertility and population structure it is necessary to define "European space".

What is Europe? There has been much discussion about whether Europe is an idea, a cultural entity, a political–economic structure or a geographical reality. According to which of these definitions we adopt, the European geographical space will have a different form, and its population will have different demographic characteristics. Even from a purely physical point of view, there are questions about the boundaries of the continent. Where does Europe begin? And where does it end? It could be suggested that Iceland is not part of Europe since it is actually closer to America than to Europe, yet historical reasons clearly support its inclusion. There are, however, other examples which are much less clear-cut. Is Turkey part of Asia or part of Europe? And what about western Siberia? Russia's expansion eastwards and the economic policies of the former Soviet Union during recent decades have meant that it is difficult to argue that the Urals should be seen as Europe's eastern boundary.

From a different perspective, and with reference to the people of Europe, how should we classify the millions of people of non-European origin, primarily Africans and Asians, who are resident in Europe, and indeed many of whom were born in Europe: Algerians, Moroccans and Tunisians in France, Asians and Afro-Caribbeans in Great Britain and Turks in Germany? And conversely how should we consider the almost 60 million Europeans under 40 years of age who have settled in Australia and the Americas? Are they still Europeans or not?

There are, therefore, considerable difficulties involved in the definition of both the space and the population of Europe. In this chapter, however, we have considered it convenient to restrict ourselves to the area that is traditionally classed as Europe, with the exception of the former USSR, but at the same time recognizing that there are huge demographic differences among the countries of Europe, and even between different areas of individual countries.

The European space is not uniform and is divided into two large areas, namely western Europe, the heart of present-day Europe, and peripheral or outer Europe.

In the past, western Europe has set the trends and the pace for demographic changes, whereas peripheral Europe has followed similar trends, but at a later date. A social space would have to be added to these two large geographical areas, comprising immigrants from other continents who live "embedded" in the most highly developed countries, although often with a low rate of integration. It is no exaggeration to say that many of the problems that the EU faces derive from these three aspects of European space. Moreover these difficulties could be increased in number if the scale is extended and the nationalism, which is appearing in many states, is taken into consideration.

The European population

Within the framework established above, one of the most important characteristics of the current European population is the trend towards stagnation and zero growth. This leads inevitably to the ageing of the population and to the formation of a block type structure which is fundamentally different from the traditional pyramid structure.

Since the eighteenth century the population of Europe has increased mainly as a result of natural growth, that is through births exceeding deaths. At the beginning of the demographic transition, the death rate fell more rapidly than the birth rate (which even rose in some areas) and this gave rise to rapid population growth (Habakkuk 1971). Subsequently, however, the birth rate has fallen more rapidly than the death rate, particularly since the Second World War, to the extent that some countries now show signs of what has been described as a second demographic transition, very different from the first (van de Kaa 1987). Rather than an increase in population, this is characterized by stabilization or even a reduction of the population, and the demographic factor responsible for this trend is the fall in the birth rate.

This has created a totally new demographic situation. In 1993 the geographical space that constitutes Europe (bearing in mind the lack of precision in the definition of the area indicated earlier) had 507 million inhabitants, that is, 40 per cent of the population of the most developed and industrialized countries of the world, which in part explains the economic and cultural importance of Europe today. However, the demographic weight of Europe in the world is gradually decreasing: Europe accounted for 17 per cent of the world population in 1850, for 16 per cent in 1950 and for slightly more than 9 per cent in 1990. Unless there is a change in this trend, Europe will probably come to account for about 6 per cent of the world population by the second decade of the next century.

The total population of Europe is currently increasing at a very slow rate;

19

Table 2.1 Average annual rate of growth
of Europe's population by 50-year periods,
1750–2000.

Years	Cumulative annual rate
1750–1800	0.4
1800–1850	0.6
1850–1900	0.7
1900–1950	0.6
1950–2000	0.6

Source: United Nations estimate.

although this rate of growth is not very much slower than that recorded in the past, it is the result of very different causes (Table 2.1).

These rates of growth are surprisingly low and consistent and would appear to contradict what is known of the demographic boom that took place in Europe from the eighteenth century on, followed by declining rates of growth (and summarized by the model of demographic transition, analyzed by Clarke (1972) and Noin (1983), among others). However, this contradiction is less evident if it is borne in mind that Europe has suffered large-scale population losses as a result of both emigration and wars over the past 150 years. Had it not been for these factors, Europe would today probably have more than 1000 million inhabitants.

The present decline in birth rates was not anticipated in earlier population projections; on the contrary, the majority of population projections made over the past decades have been higher than the actual figures. In 1965, for example, the United Nations projected that Europe (excluding the USSR) would have a population of 568 million inhabitants by the end of the century, whereas today it would appear to be certain that this figure will not be reached, despite the fact that positive population growth is still recorded in several countries.

This relatively small population raises the question as to whether Europe will be able to maintain the socio-economic and cultural pre-eminence that it has enjoyed up to now. Even a brief reflection on the likely demographic situation in Europe during the coming decades leads to the conclusion that a new demographic revolution, with its corresponding transition, is under way in Europe; the nature, extent and implications of this transition are not yet possible to calculate. We may be certain, however, that important changes will be seen in the attitudes and behaviour of the people of Europe in respect of demography. The immigration pressures which can already be felt in Europe will have an effect on these changes. There are today millions of non-Europeans who wish to become established in Europe, and they bring with them the fertility norms and attitudes of their countries of origin, which generally favour larger families. An ageing Europe, with a low rate of natural population growth, is a great temptation for non-Europeans living in countries with rapid rates of population growth and who consider they have, as a basic human right, the freedom to choose their place of residence.

Having argued that Europe is at the beginning of a new demographic transition, we should at this point, consider certain aspects of the recent revolution in European fertility.

The drop in fertility

In the traditional societies of Europe in the eighteenth century, the actual fertility rate recorded probably represented around 60 per cent of natural fertility, calculated with reference to the Hutterites (10 or 11 children per woman) or French Canadians (6 to 9 children per woman). However, these comparisons are very risky, since all simplifications lead to absurd reductionism if no account is taken of the individual characteristics of each country and, within the different countries, of each region.

Chaunu (1974) has said that the contraceptive weapon of traditional Europe was the fact that couples married late and, in some cases, there was societal pressure to limit the number of children in a marriage. These are the characteristics attributed by Bourgeois-Pichat (1965) to pre-industrial societies in which the birth rate appears to have been determined by both sociological and biological factors. In effect they are what Wrigley (1966) described as social sanctions. In traditional societies of Europe the fact that couples married late was one of the major controls on fertility, since births were closely linked to marriage. Even though in some areas there was a high rate of extra-marital births, as was the case in the countries of the Mediterranean area, even so marriages here were held within one year of the date of the birth in 90 per cent of cases. Demographers generally agree that in the highly stratified societies of the first demographic transition, the fall in fertility was a consequence of the socio-economic conditions that were created as a result of the industrial revolution. The precise mechanism by which the two interacted is uncertain. Hajnal's (1965) view that the number of births was linked to the level of economic resources has to be considered as doubtful and other explanations are needed.

The majority of theories on population growth highlight the voluntary nature of the fall in fertility that accompanied the first demographic transition from around 1870. In the most highly developed societies of Europe the attitudes and behaviour of couples came to be governed by the acceptance of new social, economic, religious and even political values that encouraged a decrease in the birth rate. This is in contrast to many less developed countries where the traditional values in favour of a high birth rate continue to predominate.

Calot has described the contemporary evolution of the fertility rate as one of the most important steps in the human adventure (cited in Levy 1989). Although this may be an exaggeration with reference to the whole of humankind, it is no exaggeration if applied exclusively to Europe. A process of accelerated change in the traditional attitudes and behaviour relating to births may be observed in the majority of European countries over the past two decades. Irrespective of the different political, religious, cultural and economic circumstances of the different countries, all have recorded the same trends. These are: a fall in the fertility rate, often very dramatic, from around 1965; stabilization of fertility at very low levels that do not guarantee generational renewal; the almost complete disappearance of families with three or more children; and gradual modification of the demographic structure with marked relative ageing of the population. All these factors allow us to speak of a second demographic transition, the nature of which we shall now explore.

The philosophy of Aristotle defined movement as "*Transitu ex uno statu in alio*", that is, transition from one state to another. This means that any type of transition consists in a process of change between the "before" and the "after". If this change is very rapid, it is generally said to be revolutionary, and in consequence, in light of the changes that are currently taking place in the population of Europe, we could indeed talk of a new demographic revolution and a new transition; a revolution and a transition which we can speculate about, but whose consequences are uncertain.

In recent decades, quantitative neopositivism has led demographic geographers to make more general use of mathematical models which have transformed demographic geography into a discipline with a high degree of autonomy within the field of geography. However, for our purposes, the processes of demographic change may be best explained on the basis of a qualitative, rather than a quantitative, analysis.

All the countries of Europe, and in particular those that form part of the European Union, have seen their fertility rates fall in the past 40 years. In France, for example, the fertility rate has fallen from 2.9 to 1.8 children per woman; in Denmark, from 2.6 to 1.5; in Germany, from 2.1 to 1.4; in Belgium, from 2.4 to 1.5; in the Netherlands, from 2.5 (3 in 1965) to 1.3 and in Spain, from 2.5 to 1.3 (Monnier 1990, Eurostat 1992). In respect of the countries of eastern Europe, the figures available for 1993 do not differ substantially from those recorded in the EU countries: Bulgaria, 1.6 children per woman; Hungary, 1.9; the Czech Republic, 1.8; Russia, 1.7; the Ukraine, 1.8; and Poland, 2.0.

A conference held in Monte Carlo in 1989 on the subject "European children and their future" devoted one session to a discussion of fertility trends in Europe. The majority of the participants in the debate linked the fall in the birth rate to socio-economic factors such as the overpopulation of the planet, inflation, the level of social welfare protection, social pressures, family incomes and the choice between two salaries or three children (conference report in Levy 1989).

Simple or multiple correlation, factorial analysis and other analytic procedures all highlight the existence of very high degrees of correlation between the fall in fertility and certain economic factors. However, these correlations, although important, are insufficient to explain the fall in fertility in Europe, since this trend affects countries with very different socio-economic, sociocultural and political characteristics. For this reason investigations must be made in other directions.

In her book *Fertility transition* (1991), Loraine Donaldson developed an idea that was held by many sociologists and economists in the eighteenth and nineteenth centuries, who believed that population growth was primarily influenced by moral and social factors. The approach of Donaldson is now widely used, not only among demographers, sociologists and geographers, but also among many economists. In the case of Spain, Higueras (1991) attributes the fall in the birth rate from 2.5 children per woman to 1.3 children in just one decade to the sociological changes that have taken place in Spain and, in particular, to the change in attitudes and norms of couples in respect of births.

In the first demographic transition the process of change began as a consequence of the gradual decrease in the mortality rate. In the transition that is now beginning the factor to be borne in mind is the fall in fertility. However, as Donaldson (1991) indicates, behind the general decrease in fertility lie many different factors, many of which cannot be submitted to accurate analysis as a result of their transitory nature. And this is because today more than ever before, the birth rate is dependent on attitudes and values that put great emphasis on personal freedom. One has only to consider the frequency with which one hears of responsible parenthood, and the importance of the debate in many European countries on the freedom to exercise control over one's own body in respect of abortion and euthanasia. The very fact that this debate is taking place indicates that there has been an extremely important change in traditional attitudes and behaviour.

Behaviour in respect of the number of children is always personal whereas attitudes towards fertility correspond to wider social values, which are part of a collective mentality formed from the combination of very different and complex elements, as shown in Table 2.2.

It is not possible to analyze in detail here the contents of each of the headings in Table 2.2, but the differences that may be observed in the rate of the decline of the birth rate in Europe are a result of varying attitudes towards fertility throughout the continent, arising from the variety of cultural and material structures within each country.

Nevertheless, what is changing in Europe is the attitude of couples towards having children and to the desired number of children, and it is for this reason that government policies to stimulate increases in the birth rate have so little success. Children are no longer considered to be economic assets, but are an element of personal fulfilment, which can be achieved with the birth of just one child.

Table 2.2 Behaviour and attitudes in respect of births and the number of children.

(a) Behaviour in respect of births and the number of children:
Cultural elements:
Legal–political structures
Socio–economic structures
Mental structures
Religious structures
(b) Attitudes in respect of births and the number of children:
Material elements:
Ethnic components
Biological components
Instrumental components

Furthermore, believers in the traditionally populist churches, such as the Catholic Church, are beginning to distinguish between love between spouses and procreation. It is, in effect, mental structures that are changing. In Spain, for example, surveys conducted on fertility show that there is hardly any difference in the use of contraceptives between practising believers and the remainder of the population. Anyone who is acquainted with the populist policies of Spanish governments up to 1975 will comprehend the huge change that has taken place in Spain in such a short time. These earlier governments used to grant prizes to large families, prizes which would be won by couples with twenty children or more. Nowadays, however, such a situation is unthinkable, and the change that has taken place in Spain is not due to socio-economic conditions, although these conditions have clearly improved, but rather to changes in mental structures, especially changes in the mentality of the younger generations (Faus-Pujol 1991). The fertility survey carried out by the National Institute of Statistics in Spain in 1977 showed that the country was still pronatalist at that date as is shown by Table 2.3.

Table 2.3 Married women who do not wish to have more children (in percentages)

Number of children	1977	1987
0	14.8	29.2
1	52.6	48.0
2	52.6	95.6

In just ten years there was a sharp fall in the desired number of children. In 1977, almost half the married women with two children wanted to have more. In 1987, very few women in this position wanted any more children. The same survey showed a fall in the number of large families, and an increase in the number of childless married women. Table 2.4 shows that the number of married women with no children has doubled between 1977 and 1987, while the percentage of women with three or more children has fallen, in spite of the fact that the survey included all married women, irrespective of age (Higueras 1991).

Table 2.4 Family size of married women (in percentages).

Number of children	1977	1987
0	5.4	11.4
1	20.0	21.2
2	34.2	34.6
3	20.3	19.2
4	10.3	7.8
5 or more	9.8	5.8

What is taking place in Spain is also taking place elsewhere in peripheral Europe (Muñoz-Perez 1987). The pace of decline in European fertility reflects the variety of geographical and social spaces within the continent, some aspects of which merit further discussion. In addition to the differences in the rate of fall of fertility that can be seen between so-called core Europe and peripheral Europe, account should also be taken of the fertility behaviour of the populations of the different social spaces of Europe.

It is currently estimated that there are some twelve million non-Europeans resident in Europe. As is shown by Noin (1991) in the *Atlas de la population mondiale (Atlas of the world population)*, the flows of emigrants towards Europe come from three areas, namely southern Asia, particularly India, the African countries of the Maghreb, and Turkey. To simplify a great deal, in Europe there are three principal areas of settlement: Great Britain in the case of the Asians, France in the case of the Maghrebians and Germany in the case of the Turks.

These immigrants come from totally different cultural environments to those of Europe and when they arrive in Europe they generally maintain the attitudes and behaviour of their countries of origin in respect of numbers of children, albeit modified to conform to the specific circumstances of the country in which they are living. In general it has been suggested that for some ethnic minority groups there may be a delay of perhaps one or even two generations for the birth rate to become adjusted to the European mean. Clarke (1972), who compared birth rates in Muslim societies in the Middle East with the European figures, found differences of ten percentage points or more between the two societies. Similarly, in Great Britain there are contrasts in the fertility of the British-born population and immigrants from Bangladesh or Pakistan, although it must also be noted that the fertility of

Indian migrants has already fallen quite markedly (OPCS 1986). In Germany there are contrasts in fertility between the native German and the Turkish community (Kane 1986).

When immigrants represent a very small minority, their attitudes and behaviour towards family size generally adapt very rapidly to the attitudes and behaviour of the surrounding community. Nevertheless, this is not so in the case of immigrants who may be counted in millions, as they tend to group together to form their own sometimes very numerous communities within which they are able to maintain their own traditions and customs, including those towards fertility. It could well be, therefore, that in the first years of the next millennium we may see an increase in the fertility rate in Europe. However, this increase will not be recorded among the native European population, but in the immigrant population. Nevertheless, ethnic barriers are increasingly less important in Europe (the ethnic clashes in the former Yugoslavia are of a different nature) and the number of mixed marriages continues to increase. Casas Torres (1982), the pioneer of population studies in Spain, has said that the twenty-first century will be the century of racially mixed populations in Europe, and this will no doubt have important effects on the attitudes of the European population in respect of family size.

Population structure

In addition to its low fertility rates, the other important characteristic feature of the European population is its high degree of ageing. In the comparative analysis of population structures the profile of the age pyramid can be used to distinguish young, mature and aged populations using a biological metaphor. But such a categorization only establishes at a very generalized level the degree of youth or ageing of geographical space and does not necessarily give a correct picture. In such a categorization Europe appears as an ageing continent.

From a quantitative point of view, the European population shows a clear process of maturing and ageing. The proportion of young people under 15 years accounted for around a third of the total population in the nineteenth century and has now declined to below 20 per cent in many European countries. Only some countries in peripheral Europe have slightly higher proportions, for example 25 per cent in Poland and Slovakia. More significantly the proportion of people aged 65 years and over, which was generally less than 5 per cent in the nineteenth century, has multiplied in the past few decades. The degree of maturity and ageing differs from one country to another, since the process is not simultaneous. In half a century, the number of elderly people in Europe has doubled and statistics show the number of elderly and very old people is still increasing.

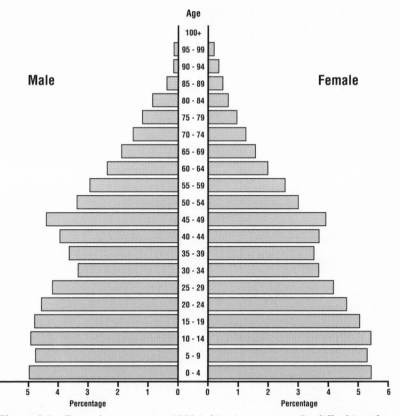

Figure 2.1 Europe's age structure 1950 (taking into account the difficulties of population data in post-war Europe).
Data source: United Nations, *Demographic yearbook.*

Guilmot (1978) estimated that from 1950 to 1970 the number of elderly people increased markedly: those aged over 75 increased by 43 per cent while those aged over 65 increased by 30 per cent, and the numbers are continuing to increase. According to Chesnais (1991a), the number of elderly people in France will increase by 0.8 per cent per annum up to the year 2000; in Germany it will increase by 1.3 per cent; in the United Kingdom by 1.5 per cent and in Italy by 1.7 per cent. Noin (1991) shows that the oldest populations in the world are in the heart of Europe; but in a world context all the populations of Europe, core and peripheral, can be classified as aged.

Although the drop in fertility began in the late nineteenth century the proportion of young people has been maintained in Europe until very recently (Fig. 2.1). In the earlier part of the century the reduction in infant mortality helped to compensate for the reduced fertility. But in the past few decades, this has no longer been sufficient, especially as improvements in mortality and hence life expectancy have

27

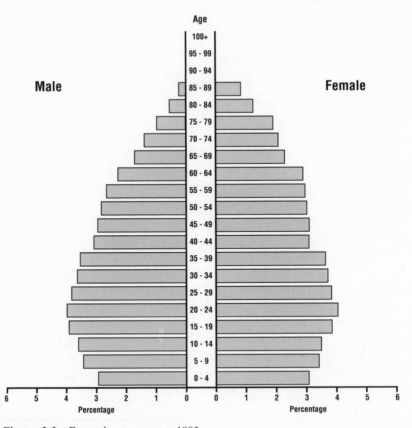

Figure 2.2 Europe's age structure 1993.
Data source: Population Reference Bureau, *Population sheet 1993.*

continued to be experienced by all age groups, including the very old (Fig. 2.2).

The first population transition was characterized by the widening of the base of the age pyramid. By contrast, the transition now under way is characterized by the widening of the summit. If the current trend continues, the pyramidal structure will disappear, giving way to a prismatic or block structure (Fig. 2.3). The change in the age structure of the population will inevitably be accompanied by great social and economic changes. The alarmists see a danger for Europe in the decline of fertility and the lengthening of life. The more optimistic claim that this is only true if the analysis is purely quantitative, but the danger disappears if other circumstances are taken into account, such as the extent of the population's participation in the socio-economic system, regardless of age, sex, and its cultural and technical level. Inevitably though we will all have to change our fundamental approach and interpretation of population growth and structure, unused as we are to such elderly populations (Amman 1985).

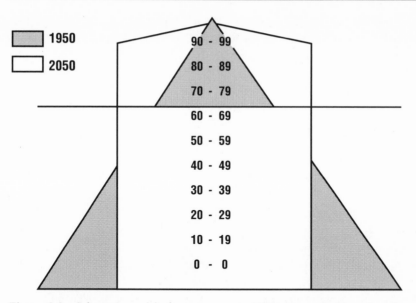

▨	**1950**
☐	**2050**

Age bands (top to bottom): 90 - 99, 80 - 89, 70 - 79, 60 - 69, 50 - 59, 40 - 49, 30 - 39, 20 - 29, 10 - 19, 0 - 0

Figure 2.3   Schematic model of age-structure in 1950 compared with projected age-structure 2050.

The study of population structure by sex and age is important on both demographic and economic grounds. On the one hand, population structure is the fundamental factor in the evolution of the population. Most demographic events (marriage, death, morbidity, migration, etc.) have a relationship to the age and sex structure. On the other hand, economists can see in the population an economic resource, the efficiency of which depends largely on its internal structure.

When we speak of young, adult and aged populations we do so because the population is viewed as a work force. Thus, the young population replaces the active population as it ages while the aged population is considered a social burden. Dependence in this way, whether global, by youth or old age, is an economic rather than a demographic indicator. The age defining the young population is set at 15 years because in most developed countries that is the approximate legal limit of compulsory education and thus access to the labour market; 65 (or in some cases 60) marks the beginning of old age, because this is the normal age of retirement. However, these age markers are increasingly regarded as inadequate.

Research into the evolution of the population structure in Spain suggests that the age defining the young population should be raised to 20 years, and that of old age to 70 or more. The reason for this is clear. In Spain, and in most European countries, the educational and vocational training process is becoming increasingly longer and rarely ends before the age of 30. At the other end of the age scale, the lengthening of life expectancy at birth, which is around 80 years, makes 65-year-olds still very far from old age.

Until relatively recently, schooling ended at 15 years in the majority of developed countries, but increasingly it is now extended to 18 years. In Spain, obligatory general education was extended recently to the age of 16, and secondary and first level professional education to the age of 18. Now, very few young people seek work before the age of 18. At the other end of the age pyramid, in some professions there is a tendency to delay retirement to beyond 65 years, so that retirement is voluntary at 65 and only obligatory at 70. This is a realistic approach, since whereas in 1960 the life expectancy of a retired person was ten years, now it is 18 years and increasing.

We can see how traditional views about age can be changed when we consider recent advances in fertility treatment. Almost all fertility is assumed to occur to women between the ages of 15 and 45 (Fig. 2.4), and yet medical advances now enable women both in their late forties and even those beyond the menopause to conceive so that maternity can be achieved by women as old as 60, as reported in Italy in July 1994. The problems raised by these techniques are not biological, but ethical, moral, psychological and social, and are issues which society has to face.

There are, therefore, many reasons why the predictions of the alarmists of the consequences of an ageing population will probably not come true. A theoretical model of Europe's projected demographic structure by the middle of the twenty-first century, in comparison to that of the middle of the twentieth century, is shown in Figure 2.3. It is a simplified, schematic model, which includes the countries of

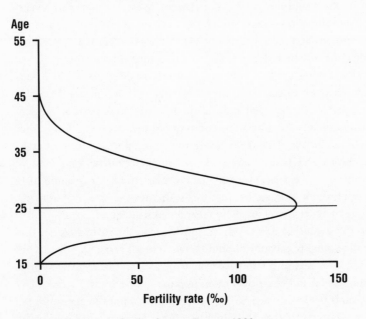

Figure 2.4 Female fertility by age, Europe 1993.
Data source: United Nations, *Demographic yearbook.*

the EU and those of central Europe. There are many imprecisions, because the information is incomplete and uneven, but it is an initial approximation of the future demographic structure of Europe.

The current trend towards increasing life expectancy and the stabilization of fertility rates at very low levels make it very likely that such a population structure will evolve in Europe. The move from the mid-twentieth century population structure to the projected mid-twenty-first century structure is the foundation of the new population transition.

Two consequences are apparent from the model of changing population structure (Fig. 2.3). The first of these is that the average age of the European population will have increased by twenty years in one century: from around 30 to 39 years in 1950 to between 50 and 59 years by the middle of the next century. There will, of course continue to be great regional differences.

The second consequence apparent from the model is the high index of ageing that Europe will experience. In 1950, 10 per cent of the total population were aged over 65; in the next century that proportion may increase to 40 per cent. By contrast, young people under sixteen years of age will represent barely 15 per cent of the population.

From a socio-economic point of view there does not seem to be any reason to question the validity of the model, assuming that current low fertility rates are maintained and life expectancy increases, as seems probable. In Spain in 1992, life expectancy at birth was almost 80 years and the National Statistics Institute forecasts a figure of 82 years by the year 2000. The population is ageing, but its efficiency is increasing as a result of new technologies. If, moreover, people's active life is lengthened and mental barriers are removed, the efficiency of the population within the socio-economic system may be maintained.

The question that arises is whether the attitude and the behaviour of the population with respect to fertility will change or whether it will remain the same as a new population structure emerges and consolidates. It is generally accepted that the current tendency for fertility to decline is unlikely to change over the next ten years. Fertility is not only a biological phenomenon, but responds to certain patterns in attitudes and behaviour which are closely related to the mental structures of our time. These patterns and behaviour affect whole generations and there is no evidence to suggest that the present young generations of possible parents are going to change their way of thinking. If fertility falls below certain limits, the renewal of generations is impossible, and sooner or later the population will disappear. This is the basis for the fear of the alarmists for whom zero growth is a real danger.

The idea that Europe could disappear as an area of population concentration is unthinkable, even though its population structure may become ever older. Neither is a change of mentality favouring higher fertility apparent among young people, which would ensure the replacement of generations. We are therefore at a cross-

31

roads with many uncertainties and it is not easy to see a way forward. There are, though, signs that events may take other directions if we include in the analysis new variables such as immigrants.

We have already referred to that particular European social space which is made up of people of non-European origin. At present there are many obstacles set up by European governments to hinder the settling of immigrants, but such policies may not be maintained for very long and certainly not for more than two or three generations. Some of the pressures are illustrated by Spain. In 1991, approximately 200000 clandestine immigrants, most of them from Africa or South America, were legalized in Spain and it is calculated that there are just as many again living illegally. In spite of the recommendations of the EU about the slowing down of immigration and the government policy in this respect, South Americans and Africans continue to arrive in Spain. The former make use of the old double-nationality agreements or come as political refugees, whereas the latter – the majority of them either from the Maghreb, Senegal or Nigeria – enter either illegally or while in transit to other European countries. The official figures refer to 200000 clandestine immigrants resident in Spain, in spite of earlier legalizations, but the non-governmental organizations speak of over half a million people who, sooner or later, might end up in some other western European country, since these organizations, one of which is the Catholic Church, protect them.

The number of non-Europeans living in Europe is estimated to double every ten years. At the beginning of this chapter we referred to this when we said that the direction of the international migratory movement was reversing. If the migrants retain the mental structures, attitudes and behaviour towards fertility of their countries of origin, they will have an important influence on the evolution of Europe's population structure in the next century. In this case the model will have to be modified by widening the base of the pyramid as a result of the fertility of the immigrants.

Conclusion

Europe is at present immersed in processes of great demographic change, the most significant aspects of which are the sharp fall in the fertility rate to levels which had previously been unheard of, together with a remarkable lengthening of life expectancy. These phenomena considered together give rise to an increasingly ageing population.

In the face of this situation, there are two contradictory views. On the one hand, those with a catastrophic vision say that Europe has begun a demographic decline from which it will be difficult to recover. On the other, there are those who feel that faced with the socio-economic changes that are taking place in the world, it is desirable to maintain population growth at zero, trusting that the lengthening of life expectancy and new technologies will somehow counterbalance what is known as the demographic deficit.

It is difficult to conceive that the European demographic deficit is an irreversible problem, at least not within the next fifty years. Nonetheless, there will have to be changes in the economic, labour and social order of Europe if it is to adjust to the new demographic conditions. With a life expectancy at birth which in the year 2000 will be around 83 years, it seems a waste of human resources that the active working life of an individual should be no more than 50 per cent of his or her lifetime.

Moreover, Europe which since the eighteenth century has been a continent of emigrants, is now undergoing the reverse process: there is great pressure from people in Third World countries which are experiencing rapid population growth who wish to settle in Europe. These are people of diverse cultures and customs who some consider do not fit easily into European cultural patterns. Those who view the decline of European fertility as a matter of concern are sure that the twenty-first century will be the century of immigrants (Casas Torres 1982). On the other hand, those who wish to maintain the current situation favour a restrictive demographic policy both in respect of fertility and immigration.

Not surprisingly, perhaps the solution lies somewhere between the two extremes. It is not certain that the fertility rate will stay at its present low levels, since it may start to rise at any time. Besides, even if restrictive immigration policies are adopted, people will still continue to come to Europe from other continents, and if they maintain their present patterns of fertility, they may well influence the development of European population structures and so modify the model projected for the twenty-first century.

CHAPTER 3

Households, families and fertility

RAY HALL

Introduction

Speculating about future demographic trends may well be extremely foolhardy, since, as Grebenik (1991) has pointed out, demographers and others who have studied population movements in the past have not been conspicuously successful in accurately foreseeing demographic developments. Nonetheless, this does not mean that examining present trends with a view to considering what future demographic scenarios may be like is without value. If we look at historical trends, patterns emerge, and we can start to appreciate the extent of change that has already taken place, as well as perhaps seeing the possibilities for future change. It may certainly help us plan for future changes and view with scepticism some analyses of current demographic structures. Ultimately though our speculations are no more than speculations.

The rapidity of recent demographic change is no more obvious than when we look at families and households within the specific context of fertility. In this chapter we examine changes in families and households over the past two decades and speculate on what the future of the family might be. In looking at families and households we also have to consider changes in fertility. Fertility inevitably intersects with wider social changes, particularly those changes affecting the role of women, and in particular the rapid increase in the number of women in paid employment.

A rising number of women in paid employment coincides with the declines in fertility apparent from the mid-1960s. By the 1970s divorce rates were rising and marriage rates falling, and during the 1980s cohabitation became a well established phenomenon. The growing popularity of cohabitation was associated to some extent with an increase in the number of births to unmarried women. At the same time there has been a rapid growth in the number of lone-parent families (Hall 1993). These various trends are not independent of each other and have to be seen as a nexus of interrelated variables. Underlying these trends are changing attitudes towards sexual behaviour and in particular the separation of sex from marriage. The nuclear family no longer dominates the household form as it did before, and especially after, the Second World War. Over the past thirty years other household types have become more numerous and households have become smaller (Hall 1988). More elderly, and increasingly, more young people, live alone; people may live in unrelated groups; family units are more complex, so that today we have an altogether more flexible and fluid family and household system than in the earlier twentieth century. The question we have to ask is, will these trends continue? Will the stable nuclear family become a rarity as people learn to expect to live in many household types during their lives and with several partners?

Changing attitudes towards sex

Households can be a single person living alone or a group of people sharing a residence, although precise definitions of households vary from country to country. Families are particular types of households and again, while definitions vary, are based on kinship usually either through blood relationships or marriage. Underpinning families are sexual relationships, so that fundamental to understanding changes in families over the past thirty years is an appreciation of changing attitudes towards sex and sexual relationships that have occurred in this period.

The 1950s are sometimes seen as a golden age for the family, a period with high marriage rates, low divorce rates and apparently stable domesticity dominating society, with sex reserved for marriage. Change began in the 1960s, a decade seen as a turning point in attitudes towards sex. For example, 1960 was the date of the first approval of the Searle formulation of the contraceptive pill by the Food and Drug Administration in the USA (McLaren 1990) and was also the date of the "*Lady Chatterley*" trial in Britain. Both events can be seen as marking turning points: the first marks the beginning of a contraceptive revolution whose impact should not be underestimated; and the second, changes in attitude towards topics that could be discussed openly for the first time. To a large extent it can be argued that the rapid changes in attitudes during the 1960s and 1970s were a response to innovations in contraceptive technology although the precise cause and effect are difficult to disentangle.

The 1960s were almost certainly not the period of wild sexual liberation sometimes imagined, and were certainly not sexually liberated by 1990s or even 1970s standards, but the decade can be seen as a turning point. Other reforms of the decade include the legalization of contraception in France in 1967 and the legalization of abortion in Britain also in 1967. These legal reforms reflect changing attitudes towards sexuality but the changes took longer than one decade. Liberalization has continued throughout the 1970s and 1980s through to the present. Similar trends

35

are evident in other European countries although generally later. Overall, perhaps the changes taken together merit being described as a sexual revolution (see Shorter 1976); the sexual changes certainly underpin the range of sociodemographic changes related to fertility and marriage.

Although it is clear that there have been dramatic changes in attitudes towards, and behaviour in, matters sexual, it is difficult to chart changing sexual attitudes since surveys are rarely identical in format either in their sampling frames or the questions asked. In 1950, a survey in England found that, not surprisingly, a high value was put on virginity before marriage. A repeat 1969 survey of people aged under 45 found that 26 per cent of married men and two-thirds of married women said they were a virgin at marriage, and most women over the age of 20 married their first sexual partner. The 1969 survey also revealed that younger age groups showed a marked shift towards a younger age at first intercourse. But remarkably, in 1969, of the under-45s questioned, 27 per cent of men and 49 per cent of women were against sexual experience for men before marriage; and 43 per cent of men and 65 per cent of women were against sexual experience for women before marriage. It also noted that premarital sex was approved of more by those born after the Second World War, that is those who were under 24 in the survey, than older people. Interestingly, women rather more than men supported a double sexual standard for men and women (Gorer 1971).

The extent of change from the late 1960s to the present is illustrated by a British survey in 1993 in which three-quarters of the respondents stated that premarital sex was not at all wrong (or rarely so). Older people expressed most disapproval about premarital sex, although not markedly so (Wellings et al. 1994). In the mid-1990s, it is estimated that 90 per cent of couples within Britain have intercourse before marriage. It is assumed that these changes in sexual attitudes and behaviour have a direct relationship with the introduction of efficient contraceptives in the 1960s, notably the pill and the IUD.

The rapidity of change between the mid-1950s and mid-1970s is shown by the 1976 Family Formation Survey, which asked married women whether they had had sexual intercourse with their husband before marriage (but did not ask about any previous relationships). The proportion of women reporting sexual intercourse with their husbands before marriage more than doubled in the twenty years between the 1956–60 and the 1971–6 marriage cohorts, from 35 per cent to 74 per cent. Three-quarters of those women marrying between 1971 and 1975 reported premarital intercourse (Dunnell 1979).

The relationship between contraception and sexual activity is not, however, a straightforward one as a study of premarital sexual activity among Scottish teenagers demonstrated (Bone 1986). The survey found that there was a rise in the proportion of girls having intercourse while single before the age of 20 from 6 per cent of those born before 1930, to 16 per cent for those born during the period 1931–5,

23 per cent for those born during the period 1941–5 and so on, to over 60 per cent for those born after 1960. There was, therefore, an increasing trend towards teenage intercourse among single women long before the introduction of the pill. An important difference, however, is that in the earlier period the age at which they married also fell, so while only 16 per cent of women born during the period 1926–30 were married by the time they were 20, this had risen to 34 per cent for women born during the period 1951–5. More teenage girls were having inter-course but they were also marrying earlier. At the same time there was a rise in premarital conceptions. For those born after 1960, while the proportion having intercourse in their teens when single has continued to rise, the age of marriage has also risen, so that only 22 per cent of women born during the period 1961–5 were married by the age of 20, and at the same time, premarital conceptions have also fallen.

These data suggest that there is now a greater use of reliable contraception (or abortion) and a different attitude to intercourse while single. While in earlier peri-ods intercourse was in anticipation of marriage, this is no longer the case and there is now a disassociation of intercourse from any idea of marriage. Comparison with other data suggests Scottish girls at this date were less likely than those in the rest of Great Britain to have intercourse while single and in their teens but the trends, nonetheless, are likely to be broadly the same. (In 1976 about 39 per cent of single girls in England and Wales aged 16–19 had had intercourse compared with only 17 per cent in Scotland at that date – Dunnell 1979.)

Changing marriage

Innovations in contraceptive technology along with the liberalization of abortion, which has occurred in almost all European countries, have had a profound impact on fertility, particularly in reducing unwanted fertility. Even more importantly, almost one hundred per cent reliable contraception backed up by the possibility of abortion has changed attitudes towards sex and marriage by separating one from the other.

Once sex and marriage are divorced then the basis of traditional marriage disap-pears. Traditional marriage, or at least marriage as formulated from the nineteenth century onwards, was based on the concept of the male breadwinner and female housewife, a domestic ideology developed primarily among the middle classes but aspired to as an ideal by the working class (Davidoff & Hall 1987). It was a bargain, it has been suggested, between the husband who provided financial support and a wife who provided sex and childcare (Luker 1984, cited in McLaren 1991: 256). Effective contraception and easily accessible abortion undercut this bargain since women are no longer risking unintentional pregnancy by having sex outside marriage.

Efficient contraception, from the 1960s in particular, meant women could plan with confidence whether and when they would become pregnant and also assume they would not have later, unintentional pregnancies. These changes coincided with a rapid growth in paid employment among women and childbearing could, for some women at least, be slotted into their career patterns.

The rise in the proportion of married women in paid work is one of the most notable changes to families since the 1950s. In Britain for example, 30 per cent of married women worked in 1951; by 1991, 62 per cent of married women with dependent children were in full or part-time work. Careers have become increasingly important for women. This was exemplified in a recent survey in Britain by the market research organization Mintel, which found that women as much as men see their self-esteem bound up with their job and working. Women who are mothers place as much importance on their jobs as do men: 34 per cent of working mothers with children aged 11 to 16 felt that their job performance was central to how they feel about themselves – the same figure as men. Among working mothers, 47 per cent would carry on working even if they did not need the money, compared with 46 per cent of men (*Independent*, 28 July 1993). The economic base of traditional marriage has undergone fundamental change and the way women view themselves has changed equally profoundly. These changes are reflected in changing marriage rates.

CHANGING MARRIAGE RATES AND PATTERNS

From the 1970s onwards we can see a decline in marriage rates taking place in all European countries along with a rising age of marriage. In some ways, it can be argued, this was a return to an earlier pattern of late age of marriage traditional in Europe up until some time around the Second World War (Hajnal 1965). Along with late age of marriage, the traditional European marriage pattern incorporated high proportions of the population who never married – and remained celibate as demonstrated by low rates of non-marital births – together with moderate levels of fertility. Sexual self-control was an essential component of this pattern if young people could not marry until sometime in their later twenties. There may have been a variety of forms of marriage, and sexual intercourse may have been allowed after betrothal and preceding the wedding, but illegitimacy rates remained low. In England and Wales, for example, illegitimacy rose to around 7 per cent in the mid-nineteenth century, falling again in the later nineteenth century, and rising once more in the twentieth century to reach around 9 per cent in the mid-1940s, before falling to around 5 per cent in the 1950s (Laslett 1980). Current rates of non-marital fertility therefore show a very marked discontinuity with anything in the past.

The pattern of marriage rates and age of marriage was relatively uniform throughout Europe in the decades after the Second World War. Marriage rates were high and young age of marriage was the norm everywhere, although marriage tended to be earlier in eastern than western Europe. From a high point in the 1960s, the 1970s onwards have seen an overall decline in marriage rates, and an increase in the average age of marriage. At the same time, there have been increasing contrasts within Europe with the earlier pattern persisting in eastern Europe, while in western Europe and especially northern Europe, proportions marrying have declined and age of marriage increased. These trends have been most pronounced in Sweden where the marriage rate per 1000 in 1991 was only 4.2 (Council of Europe 1993). In eastern Europe since the collapse of the Soviet Empire, marriage rates have declined somewhat but still remain higher than in the rest of Europe. There are contrasts in western Europe too, with southern and northern Europe having the lowest marriage rates and Great Britain, the Netherlands, Belgium and Germany having intermediate rates. In most countries, the mean age of marriage has increased by two or three years compared with the 1960s.

There are a variety of explanations for the decline in marriage rates, one of which is related to improved contraception. The importance of improved contraception in explaining the decline in marriage rates is shown by the decline in the number of marriages occurring because of a conception. In England and Wales, for example, one study showed that 41 per cent of the decline in marriages between 1971 and 1983 could be explained by the decline in such "forced" marriages. For France, for a similar period, the figure was 37 per cent; for West Germany between 1968 and 1983, 68 per cent; whereas for Denmark between 1972 and 1983, nearly all the decline in the marriage rate – 93 per cent – was explained in this way (Bourgeois-Pichat 1987: 19).

Marriage is still taking place but at older ages. What has changed is that marriage no longer marks a distinct break in life-style or household forms as in the past, nor acts as a regulator of sexual relationships. It has been suggested that increased individualism has developed since the 1960s with much greater emphasis on individual autonomy (Lesthaeghe & Surkyn 1988, van de Kaa 1987). This has resulted in weaker social ties and partnerships based on equality rather than the asymmetric power structures of traditional marriages. As a result there has been a growth of alternative life-styles including alternatives to conventional marriage. At the same time, couples have been more willing to end unsatisfactory relationships by divorce. It seems unlikely that there will be a return to the marriage patterns of the past anywhere in Europe, unless there were economic benefits in being legally married. Rather we are likely to see a spread of the northern European pattern to those regions of Europe where cohabitation is still less frequent.

The rise in divorce

Coincident with a decline in marriage has been an increase in divorce. Divorce rates have risen with varying intensity in different European countries. To some extent the differences can be explained in terms of divorce legislation. Rates are low – between one-quarter and one-third those elsewhere – in countries where divorce is difficult or expensive to obtain, mainly southern European countries and Ireland (where it is prohibited). Even after some liberalization of the law in both Spain and Italy, rates are still substantially lower than in northern and western Europe. Overall the probability of a marriage ending in divorce has risen rapidly since the 1960s. In the 1950s only about 10 per cent of marriages ended in divorce in western and northern Europe. Today in Sweden, Denmark, Norway and the UK something like 40 to 45 per cent of marriages end in divorce, and in other northern and western European countries, between 30 and 40 per cent. There has, however, been some stabilization of divorce rates in the 1980s, which might reflect the growing unpopularity of marriage and the growth in cohabitation rather than being a result of more stable marital relationships.

Attitudes towards divorce have also changed and divorce is now an accepted part of European society. The presence of children is not regarded as a possible barrier to divorce, so much so that in Britain a recent survey found that six out of ten people thought it better for the children if an unhappy marriage ended (National Children's Bureau 1993). The instability of marital relationships is now regarded as the norm. But we should remember that marital instability is nothing new: in Britain, for example, marriages today still last about the same time as in the nineteenth century, only in the past they were dissolved by death and today by divorce (Anderson 1983). As life expectancies continue to improve, perhaps it is too much to expect that anything more than about half of all marriages will survive until the death of one of the partners in their late seventies or eighties. Similarly, we should not be surprised if marriage as an institution continues to become weaker, although this does not mean that it will disappear completely.

Rise in cohabitation

The fall in marriage rates experienced in the 1970s was accompanied by a rise in cohabitation, a very marked break with social conventions in the earlier part of the twentieth century when living together outside marriage in Britain as least was often described as "living in sin". Again parallels with the past can be found: in eighteenth-century Britain a range of informal marriage practices have been recorded, including "besom weddings" and "living tally" (McRae 1993), or in

Scandinavia where betrothal marked the beginning of sexual relations. Cohabitation started to increase in the 1960s in Sweden and Denmark, and in the 1970s in other northern and western European countries. Cohabitation appears to be rare in southern and eastern European countries, although by its very nature, data on all cohabitation are elusive and are dependent on social surveys. Today, therefore, in many European countries large proportions of couples cohabit before marrying and in Sweden, for example, often choose not to marry at all. Since these are relatively new developments the full implications for future family and household structures are not at all clear.

Cohabitation at present is a phenomenon of younger age groups, particularly people in their twenties, and has been termed nubile cohabitation by Kiernan & Estaugh (1993). There are variations in cohabitation between countries and within countries. In some cases it is a childless prelude to marriage, in others there is childbearing while cohabiting although often marriage takes place subsequently, and in some cases it can be seen as an alternative to marriage. Cohabitation has developed furthest in Sweden and Denmark where there is little distinction between marriage and cohabitation and where living arrangements are seen as essentially a private matter (Kiernan & Estaugh 1993).

Increasing proportions of couples cohabit before marriage. In France 20 per cent of all unions began outside marriage in 1968 while by 1985 this had increased to 65 per cent, so that within a period of twelve years there was an almost complete change in norms (Leridon & Villeneuve-Gokalp 1988). Interestingly, the almost linear increase between 1968 and 1982 appears to have stopped in the past few years, suggesting that perhaps premarital cohabitation has reached an upper limit (Leridon 1990). In West Germany 20 per cent of women who married before 1974 had experienced premarital cohabitation; this figure increased to 35 per cent among those married 1977–8 (Hopflinger 1985). In Great Britain 34 per cent of those married during the period 1980–84 had cohabited compared with 50 per cent of those marrying during the period 1985–8. It has been suggested that by the year 2000, 80 per cent of couples will cohabit before marriage – if of course, they decide to marry (Dormor 1992).

Survey data from France give some insight into who cohabits and perhaps why. Cohabitation appears to be more likely among those who had experienced the divorce of their parents and were brought up by only one parent. Daughters of working mothers were also much more likely to cohabit in order to maintain their independence. As cohabitation has become more common so the social acceptability of cohabitation has increased, which in turn has reduced the pressure on couples to marry (Leridon & Villeneuve-Gokalp 1988). These data suggest that cohabitation will continue to increase in prevalence, not least because more children will experience the divorce of their parents.

At present cohabitation is most prevalent among young people, but, as has already taken place in Sweden, it is likely to increase in prevalence among older age groups. If Sweden is the European precursor of family trends then cohabitation will exist alongside marriage with little or no differentiation between the two. In general cohabiting relationships tend to be impermanent or lead to marriage after two or three years of cohabitation. Some couples may well see it as a strategy to avoid the risk of divorce, especially if they have previous experience of divorce, either themselves or of their parents.

It seems unlikely that there will be a return to the relatively stable relationship patterns of the past, even if relationships were stable in the past. In this context it is important to remember that when divorce was difficult or impossible for the majority of the population, unknown numbers of marriages ended through desertion. The future is likely to see a continuation of present trends. Partnerships are likely to become increasingly informal and perhaps begin without the assumption that they will last a lifetime. Perhaps in any case, an expectation that the majority of relationships will last a lifetime is unrealistic for couples who may begin relationships in their late teens or early twenties and who have life expectancies of at least 80 years. This is especially likely since now relationships are more concerned with companionship rather than based on the need for economic survival.

Couples today have high expectations of their partners so that there is considerable scope for disappointment. These high expectations contrast with what appears to have been the rather different expectations from marriage for earlier generations. One 1955 survey found that couples saw success in marriage in terms of the efficient fulfilment of the roles of breadwinner and homemaker (Gorer 1971). Moreover, women in the past rarely had any choice but to stay in unsatisfactory relationships. Today many more women are economically independent so that they can decide whether to stay in a relationship much more easily than was the case in the past. One could also argue that in the future, as fewer marriages are contracted, then these may experience lower rates of breakdown as people adjust to the new realities and perhaps only marry at older ages. It is likely that these changes in marriage, cohabitation and life-styles already current in the north and west of Europe will spread to other regions of the south and east. The scanty data available suggest that urban areas in both Spain and Italy are experiencing the beginnings of such changes (Del Campo 1990, Golini 1987). Overall in Europe, it seems likely that increasing proportions of people will experience a series of relationships lasting months or years, either formally recognized by marriage or informal cohabitation. In this way serial monogamy may well become the norm.

Non-marital births

As marriage rates have declined and divorce rates risen so too has the proportion of non-marital births risen. This, more than any other indicator, illustrates the extent of change that has already taken place, as well as providing an indication as to how changes might proceed in the future. Marriage is no longer the necessary precursor for bearing children. Women who have inadvertently conceived may choose not to continue the pregnancy or may decide to bear the child without feeling it necessary to marry the father. Some single women now make a conscious choice to have a child, again with no intention of marrying or even cohabiting with the father.

The rise in extramarital births began in the 1970s: for the countries of the European Union for example, from around 5 per cent of all births to 16 per cent in 1988. Scandinavian countries have the highest proportions: Sweden, where 52 per cent of all births were extramarital in 1990, has the highest rate closely followed by Denmark (45%). In the UK and France rates are approaching one-third of all births. Numbers of extramarital births in southern European countries remain very low, although even these countries have seen some increases in numbers. Many of the non-marital births are to cohabiting couples: in England and Wales 54 per cent of non-marital births were registered by both parents living at the same address, while a further 20 per cent were jointly registered although the parents did not register the same address. In Sweden and Denmark the majority of such births are to cohabiting couples.

In the USA where there has been a similar rise in births to unmarried women (over 27% of all births by 1989), one study reported that two-thirds of the never-married women had not intended to become pregnant compared with about one-third of all married women. Even so, the same study showed that for some single women, particularly older women, single parenthood was often a deliberate choice (Ahlburg & De Vita 1992)

Some ethnic minority populations are likely to remain exceptions to these trends and their household structures are likely to continue to be traditional in character – particularly among Moslem groups. A recent study of household and family formation among different ethnic groups in Great Britain from data in the 1991 census showed that Asian women were much less likely to cohabit than white women. Similarly births outside marriage are very uncommon for Asian women and it is within the context of marriage that Asian women are likely to continue to bring up their children (Heath & Dale 1994).

Trends in fertility

The increase in non-marital births has to be seen against a backcloth of declining fertility rates everywhere in western Europe from the mid-1960s onwards. The decline was particularly rapid in northern Europe up until the mid- to late-1970s stabilizing in the 1980s and even rising in some Scandinavian countries, most notably Sweden. In southern Europe, the fertility decline took place somewhat later in the 1970s and has continued up to the present day. Ireland, too, which for most of the period since 1945 has had one of the highest fertility rates in Europe, has experienced rapid decline since 1980 and by 1992 fertility was at replacement level. Spain now has the lowest fertility in Europe with a provisional total fertility rate (TFR) of 1.23 in 1992, closely followed by Italy (which had previously held that position since 1986) with a TFR of 1.25. The ban on contraception in Spain was lifted as recently as 1979, and a major decline in fertility took place in the 1980s. These are much lower rates than are current in much of northern Europe; indeed, in Sweden, fertility has risen above replacement level in 1990 and 1991 for the first time in twenty years.

By 1991, then, with the exception of Sweden, births were lower everywhere in western Europe than in 1970. In Italy and Spain they had fallen to about 60 per cent of the 1970 figure while in countries such as the UK, France or the former West Germany to just below 90 per cent of the 1970 figure. What is interesting is to compare the composition of these births. In 1970, proportions of non-marital births were low everywhere apart from Sweden; but by 1991 some countries such as the UK and France had seen very large increases in non-marital births, while in others the proportions were still low, although they had risen over the past twenty years. So women everywhere in Europe are choosing to have fewer children but the circumstances in which they have these children are very different. In Sweden, which had a larger number of births in 1991 compared with 1970, nearly half of those births were non-marital compared with only 18 per cent in 1970. By contrast, in Italy, where fertility has fallen to very low levels, the vast majority of these births are within marriage (Table 3.1). In Ireland, on the other hand, non-marital births have increased as a proportion of total births as the birth rate has declined: in 1980 they comprised 1 in 20 of all births rising to 1 in 6 by 1992.

The age at which women are choosing to have children is also rising, more women are having their first child between the ages of 25 and 29 years and increasing numbers between 30 and 34 years. Increasing numbers of women are choosing to have only one child, especially in southern Europe, and there also appears to be a rise in numbers choosing to have no children.

What are future fertility levels likely to be? Fertility is the most unpredictable of the demographic variables and even if we knew exactly what motivated couples to have children and when, and if we knew how the economy would develop, it

Table 3.1 Non-marital births as a percentage of total births 1970 and 1991 for selected European countries.

Country	1970 births	% non-marital	1991 births	% non-marital
UK	903 907	8.0	792 502	29.6
Sweden	110 150	18.4	123 561	48.2
West Germany	810 808	5.5	721 251	10.5
Italy	901 472	2.2	559 390	6.6
Spain	662 433	1.4	386 509	9.4
France	850 381	6.8	759 000	30.1
Ireland	64 382	2.7	52 690	16.6

Source: Council of Europe, 1994, table T3.2.

would still be almost impossible to predict. As Cruijsen (1991: 2) has said: "our strongly industrialized and institutionalized society is built upon the principle 'change'". The ultimate proportion of childless women is likely to rise from 10 per cent among females born in 1945 to 18 per cent among those born in 1955, and it may continue to rise. Cruijsen (1991) has suggested two likely future fertility scenarios depending upon governments' support for child-rearing. In one scenario, fertility will continue to decline as a result of the full emancipation of women and the growth of egalitarian gender roles unless conditions are created which allow women to combine motherhood with a career. If no action is taken to make it easier to combine a career and motherhood then childlessness may rise to 25 per cent, and those who do have children will have only one, or at most, two. In this scenario, completed fertility will be around 1.5 or less. In the second scenario, if support is given to child-rearing, which would entail extending childcare facilities, increasing family allowances and male partners actively participating in child-rearing and domestic work, then fertility could rise to around replacement rates – 2.0 or 2.2. But in either scenario, the average age of motherhood will continue to increase as women wait to have children, either to see if the child-rearing situation is improving or to ensure that both partners are secure in their jobs and that they have an established network of care before embarking on childbearing.

Sweden has already shown that a positive public policy towards helping women combine work and motherhood can boost the birth rate to above replacement levels, since Sweden combines high female labour force participation rates with high fertility (Sundstrom & Stafford 1992) (although the current Swedish birth rate is partly a result of the timing of births and is likely to fall slightly). Nonetheless, fewer Swedes are remaining childless and more are having two or even three children than in most European countries.

Keyfitz (1993) has argued that in a society in which women are equal to men then fertility will inevitably fall to very low levels. The domination of women and a male breadwinner ideology are a necessary part of a culture to produce high fertility. If women are free to choose how they wish to live then the majority will choose to have very few children, and many may well choose to have none. So truly effective equality, which may be achieved in the next century, he argues, will result in even lower fertility than that of today.

Bourgeois-Pichat (1989) has spoken of a demographic implosion in the twenty-first century as the birth rate remains well below replacement level and population decline inevitably occurs – as the UN already projects it will towards the end of the second decade of the twenty-first century in Europe (UN 1991). Biology, though, may intervene. In his 1989 paper, Bourgeois-Pichat speculated on the possible mastery of the menopause and what this might imply for fertility in the future. Already, we now know that this is possible, having witnessed so-called designer babies, born to post-menopausal women through egg donations. Other developments in the field of reproductive biology are also no doubt only too likely, so that it may be that older women, who have satisfied their desires for careers, may be able to choose to have children in their late forties or fifties, or even, as has already happened, in their sixties. This may help, even if only slightly, to increase the birth rate. Many people regard such late fertility as undesirable, but attitudes towards late fertility may well change as it becomes more common. Older mothers may have more time to devote to their children, and time is a commodity in increasingly short supply among younger women. Given that female life expectancies are already around 80 years and rising there is no rational reason why late fertility should be discouraged.

There is also discussion about future trends in ethnic minority fertility. At present in west European states the groups with the highest fertility are the more recently settled communities of migrants with youthful age structures and the fertility norms of their place of origin: Turks in Germany, Bangladeshis and Pakistanis in the UK and Algerians in France (White 1993a). There are two views about future trends. One is that fertility will decline with length of residence among future generations – as has already happened among West Indians and to a lesser extent Indians in the UK for example. The other is that the minority effect, reinforced by cultural and religious differences, may sustain somewhat higher fertility levels (Coleman & Salt 1993).

A second demographic transition

The various sociodemographic changes discussed in this chapter – declining fertility, declining marriage, rise in cohabitation and rise in divorce – are all closely inter-related and can be summarized by the phrase "a second demographic transition" (van de Kaa 1987), a stage in demographic evolution that will lead inevitably to population decline in Europe. Van de Kaa has described the norms and attitudes that highlight the contrasts between the first and second demographic transitions as altruistic and individualistic: "the first transition to low fertility was dominated by concerns for family and offspring, but the second emphasizes the rights and self-fulfilment of individuals" (p. 5). Morsa (1978) discussed the changes in the family which have resulted in low fertility, and again the emphasis in his discussion was on a move away from a traditional family devoted to collective goals to the family as a vehicle for individualism. Inevitably, a family subordinated to the aims of its individual members is precarious. "Founded because it is expected to bring enrichment, the family is allowed to break up if these hopes are disappointed" (p. 57).

In understanding the sociodemographic changes, the emphasis must be on changes in attitudes that have emerged among the generations born since the end of the Second World War. Such attitudinal change may be associated with a variety of other trends including secularization, egalitarianism and emancipation, for example, with a particular emphasis on individual self-fulfilment. Changing attitudes and expectations among women in particular are driving much of the household change. Large numbers of children are not compatible with the norms of individualism, and for women in particular, the opportunity costs associated with childbearing are considerable. It is therefore unsurprising if women are increasingly reluctant to have more than one or two children, especially as all surveys of domestic work show that women still have the major responsibility for such unpaid work so that they face the double burden of housework and paid work. Well educated women in highly paid jobs are unlikely to continue to tolerate such a situation. Either men change or women will live alone or in more egalitarian households. Cohabiting households, for example, appear to manage to divide household duties more evenly than married households (Matthiessen 1991: 416).

Other household types

The emphasis in this chapter has been on family households, but it should also be noted that households in general are changing. Non-family households are increasing in importance especially as more people live alone: for example, 27 per cent of all households in France and Great Britain were one-person households by 1991 compared with 20 per cent and 18 per cent respectively in 1970. This growth in people living alone is a result of a range of factors including the increasing proportions of elderly in the population, as well as increasing numbers of younger people choosing to live alone either after leaving the parental home or as a result of the dissolution of a marriage or a cohabiting relationship. To some extent it can be seen as an indicator of the increasing number of transition periods in an individual's life. And again this trend can be related to wider attitudinal changes. A person's identity

47

is becoming less associated with being part of a wider unit but as an individual with an independent life-style. The young and highly educated, living in large cities, are among the first to adopt this life-style. Amongst young men aged 25–29 in Britain the proportions living alone doubled from 5.6 per cent in 1981 to 10.1 per cent in 1991 (Murphy & Berrington 1993). Younger people are also increasingly likely to live with other unrelated people. Older people too want to determine their lives independently for as long as is possible. The proportions of over 65s in the population are projected to continue to increase next century, to about 20 per cent or more of the population in most European countries. As life expectancies increase there will be lower proportions widowed in the older age groups, although inevitably amongst women in the oldest age groups widowhood will be high. But as divorce rates also increase, more older people will be living alone as a result of divorce.

Overall, we can predict that the variability of households throughout Europe will increase next century. At the same time households are likely to become more impermanent as moves into and out of different household types increase in frequency. Children in particular are likely to live in a variety of household types as the marriages of their parents break down, and are replaced by other relationships or other living arrangements. The divorce rate among second marriages is higher than among first, one reason for which has been suggested as the presence of stepchildren (Hoffmann-Nowotny & Fux 1991). Marital instability is likely in turn to shape children's attitudes to desirable family forms when they mature and, it has been suggested, produces an intergenerational transmission of marital instability (Pope & Mueller 1976).

Policy implications

Increasing numbers of households and changing household types have a variety of political and social policy implications. With declining population totals (through below replacement level fertility) declining housing demand might be expected. In fact, rising numbers of households mean continuing pressure on housing and rising rather than falling demand, especially for housing suitable for single-person rather than multi-person households.

As households with dependent children become a smaller proportion of the total so they may need more state support, in particular, those families with only one or even no wage-earner. Lone parent households are one such group. Within Britain for example, lone parenthood has produced considerable political and moral debate especially as the increase in the numbers of lone parent families is viewed as a considerable burden on the state. The living arrangements of the very old will also increasingly pose questions for social policy. The process of individuation and emphasis on self-fulfilment does raise questions as to who pays for an individual's desire for freedom. For example, does the state have the responsibility for paying for the support of children after the break-up of a family? How can the family remain self-sufficient with little or no support from the state, as has been the case in the past, if the break-up of families is high? In Britain the attempt to make fathers pay realistic maintenance for their children after divorce and the creation of the Child Support Agency has led to large numbers of fathers complaining about excessive payments. And the question remains: who pays for the freedom to live as we wish?

Conclusion

We do not know whether the trends outlined in this chapter will continue so that households will become ever smaller with ever larger proportions of people living on their own. Already in inner London, 38 per cent of households are single-person and projections suggest that single-person households will increase to 35 per cent of all households nationally by 2011. Will there be ever increasing instability of families and households as people continue to search for personal self-fulfilment? Or might there be a backlash and an establishment of less egalitarian values, which might bring a rising birth rate and more stability to household structures? Nothing is inevitable and yet it seems unlikely from the vantage point of the mid-1990s that women in particular will return to the values and behaviour of an earlier age and, in particular, large families.

There has been much discussion about the future of the household, and whether or not we are seeing the destruction of the family. It is probably easier to acknowledge that the family is changing but not necessarily being destroyed. It is likely though, that if divorce rates in particular continue to rise then the parent–child, and particularly the mother–child relationship, will become the most permanent and enduring relationship. Other adult relationships will continue to become more impermanent and, in particular, serial monogamy will replace lifetime marriage.

Other factors are also likely to have an impact on the shape of the family and household in the future. For example, will AIDS make a difference to sexual morality making people more cautious in their sexual relationships and perhaps slowing trends towards serial monogamy? How will developments in biological research affect the family? How will society react to the so-called designer babies, born after the menopause to much older mothers? Will other scientific advances change our attitudes to fertility so that we will accept a range of other interventions including determining the sex and other characteristics of a child. Timing of the conception of a child may become much more certain and at the same time, contraception may become totally reliable. What, in fact, would be the impact on fertility of a hundred per cent reliable and easy method of contraception? Or are we likely to resist such changes and remain conservative in this field of human behaviour?

The trends that we have observed since the 1960s may only be temporary and fertility may of course rise once more. The difficulty of making predictions is demonstrated by the discussions of the 1930s and the widespread fear of declining birth rates; such fears were shown to be unfounded by the rapid rise in fertility after 1945. The difference today is that now we have a more highly educated population than ever before in Europe, and women as well as men expect to take their place in the world of work; at the same time men are expected to take a more active role in the world of home.

In a sense it is how far these role interchanges will go that will determine the future of fertility and the family. This is difficult to predict, especially as so far there is no real sign of fundamental changes in male attitudes. But to reiterate the theme that runs through the chapter: it is difficult to imagine a reverse or retreat of women back into the home and therefore it is difficult to envisage higher order births ever becoming widespread again. Future changes in Europe's population will arise from changes in the structure and organization of the family and particularly how the tensions between women's various roles and the family can be resolved.

What we can be certain of is that although some people may deplore the disappearance of so-called traditional values and there may be nostalgic yearning for a return to so-called basics, the family and household system of the 1950s has disappeared for ever and whatever the twenty-first century might bring it will certainly not be the family of the early post-Second World War period.

CHAPTER 4

The labour market aspects of population change in the 1990s

ANNE E. GREEN & DAVID OWEN

Introduction

This chapter identifies some of the main labour supply implications of the changing age structure of the population and migration trends within the United Kingdom and western Europe in the 1990s, placing them in the context of likely changes in the character of labour demand over the same period. As the populations of many western European countries remain relatively stable in size, increasing attention has been focused on two aspects of population change: first, changes in age structure; and secondly, changes in spatial distribution. The foremost features of the changing age structure of the population focused on in this chapter are the increase in the number of elderly people, and the decline in the number of school-leavers. With regard to changes in spatial distribution, the emphasis is on macro-scale trends; notably, the differential growth and decline of urban and rural areas, and of core and peripheral regions. These changes in age structure and spatial distribution have implications for the size and composition of the population of working age in different areas.

So much for changes in labour supply. Patterns of labour demand are also projected to change in important ways by industry and by occupation. At a broad level of generalization, the key changes in employment by industry are continuing job losses in agriculture and manufacturing, and the expansion of jobs in services. The main occupational trends are a forecast increase in professional, associate professional, and managerial & administrative occupations, alongside projected job loss in skilled and unskilled manual occupational categories (IER 1991, 1993).

The interactions between labour supply and demand changes have uneven consequences for unemployment, participation rates and migration – by population subgroup and across space (Green & Owen 1991). A broad distinction can be made between long-run demographic/labour market interactions, and the short-run labour market consequences of demographic change (de Jouvenal 1989, Lindley

1990). Here the main emphasis is on the implications of long-run trends from a strategic perspective, rather than short-term labour market effects – which may achieve crisis proportions in some local areas at some time periods. Population fluctuations can have a variety of effects on a socio-economic system, depending not just on the nature and time-scale of the changes giving rise to those fluctuations, but also upon their conjunction with other trends and events. Whether the key demographic changes highlighted in this chapter will be problematical depends on a number of other features of the socio-economic system: the context of population changes is of the utmost importance.

The remainder of the chapter is divided into seven sections. The first contains a summary of the main features of demographic projections over the medium-term, with a particular emphasis on the changing size and composition of the working age population relative to other age groups. The focus shifts explicitly to labour supply in the second section. The third section deals in more detail with changes in labour demand, highlighting the growth of new forms of employment and identifying the areas most likely to benefit and lose out in the face of such changes. Three key labour market outcomes resulting from mismatches in the interaction of labour supply and demand – unemployment, inactivity and migration – form the focus of interest in the fourth section; while in the fifth section there are reviews of selected key labour market issues – the ageing of the labour force, possible skill shortages, education and training, and subgroups (older workers and women). Some of the associated strategic policy issues arising from the trends identified in these five substantive sections are reviewed in the sixth section, while the key conclusions of the analyses are presented in the seventh section.

Key features of demographic change

Projections suggest that the population of western Europe (here defined as the pre-1995 12 members of the European Union) is likely to remain stable during the 1990s, as a result of the continuation of the long-term trend of low birth rates in northern regions, combined with rapidly falling birth rates in the southern regions and Ireland (CEC 1989–93). Over the period 2000 to 2015 population decline at a rate of 0.25 per cent per annum is expected. However, there are differences between countries in the timing, scale and direction of population changes. In what was West Germany, Luxembourg, Belgium and Denmark a decline throughout the period to 2015 is expected. By contrast in Greece, Italy and Portugal the picture is one of relative stability to 2000 and decline thereafter. Growth in France, Spain, the Netherlands and the UK from 1990 to 2000 is projected as the forerunner to stagnation thereafter; whereas in Ireland growth is forecast throughout the period from

1990 to 2015 (CEC 1991a). By contrast with this overall picture of stability in terms of overall size, outside western Europe population growth is projected to continue. This growth is perhaps best exemplified by a projected doubling in the population of the Maghreb countries by 2025 (King 1991).

However, this overall stability in population size in western Europe disguises marked changes in age structure. The proportion of the population aged over 65 years is set to rise from 13 per cent to 19 per cent over the period from 1990 to 2015, whereas the share of the total population aged under 15 years is expected to fall from 20 per cent to 15 per cent over the same period.

These ages mark the traditional cut-off points for defining the working age population. The change in working age population in an area over a specified period depends on the balance of endogenous population growth/decline and inward/outward migration. In western Europe as a whole, the working age population increased by around 1 per cent in the second half of the 1980s. Increases tended to be higher in the southern regions than in the northern regions. In the Paris region and South East England, there was little growth in the working age population; in Greece, southern Italy and much of Spain and Portugal the increase was more than 2 per cent. Although areas with younger populations are projected to see a considerable decline in the number of children over the longer-term, in the medium-term the higher fertility rates in the past in these areas will maintain the relatively rapid population growth in the working age population. In the past, net migration outflows have tended to offset relatively high population growth in less developed regions.

A useful device for measuring size of the working-age population relative to that of children and older people is the total dependency ratio. This statistic measures the number of people aged under 15 years and over 64 years, relative to the number of people in the working age groups. For western Europe, the total dependency ratio attains its lowest level in the early 1990s: with total dependency ratios of less than 40 per cent in parts of northern Italy, Germany and the Netherlands, but over 50 per cent in many regions of Spain, Portugal, Greece, Ireland, South West England, Wales and large parts of France. By 2025 dependency ratios rise to a level akin to those experienced in 1980. However, the old dependency ratio (the number of people aged over 64 years relative to the number of people aged 15–64 years) increases from the mid-1980s onwards, while the young dependency ratio (the number of people aged under fifteen years relative to the number of people aged 15–64 years) declines slowly (see Fig. 4.1). By 2010 the old dependency ratio exceeds the young dependency ratio. This increase in the old relative to the young dependency ratio is common to all western European countries.

This simple analysis of dependency ratios indicates an ageing of the population. This ageing is particularly severe in Germany, northern Italy and much of the Netherlands and Belgium. There are several important labour market and social policy

53

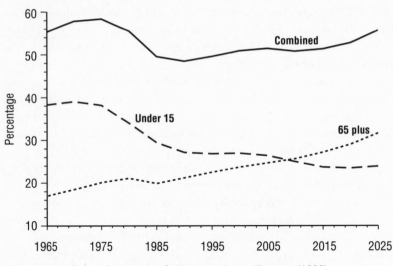

Figure 4.1   Dependency ratios for Europe. *Source:* Eurostat (1988).

themes emerging relating to the long-run financial, economic and social implications of the continuing trend of relatively low birth rates and associated ageing of the population. These include the need to finance future pensions and care for the elderly (Ermisch 1990, Hills 1993), the treatment of retirement options (Rein & Jacobs 1993), the possible waste and wasting of skills and experience of older workers, possible skill shortages and associated requirements for flexible personnel policies to meet demand, and issues concerned with equity – between population subgroups and areas. For example, the issue of the sustainability of intergenerational solidarity arises under circumstances where the financial burden of providing pensions and care for the elderly becomes much heavier because of the rising ratio of retired people to labour force participants, and the rising expectations of older people as to the standard of living they wish to maintain (de Jouvenal 1989). However, as well as the association of ageing with certain personal services, provision of which is dominated by the public sector (e.g. health, community care), changes in age composition also affect the profile of personal and household consumption, investment and saving – and this in turn has an impact upon labour demand.

Implications for labour supply

The size and structure of the labour force in an area depend on two factors: first, the size and structure of the total population of working age; and secondly, the activity/participation rates in each sex and age group (Department of Employment

1991). Between 1975 and 1985 the labour force of France, West Germany, the Netherlands, Belgium, Luxembourg, Denmark, the United Kingdom and Ireland increased by 13 million. For males, the population effect (measuring the impact of demographic changes) accounted for an increase of 5.5 million, but a negative participation rate effect (measuring the changing propensity of population subgroups – in this case a reduction in the propensity of males – to enter the labour market) acted to reduce the economically active population by 2.4 million, with the result that the actual increase in the male labour force was only 3.1 million. For females, both the population effect and the participation rate effect were positive – the increases of 3.2 million and 6.6 million, respectively, bringing the increase in the number of economically active women to 9.8 million. Relative to the 1975 labour force, there were some marked variations in the size of the population and participation rate effects between countries. In Italy and Denmark the participation rate effect was positive for males, contrary to the general trend, whereas in Belgium, Luxembourg, Ireland, the Netherlands and France the participation rate component served to reduce the male labour force by more than 5 per cent. The population effect was positive in all instances – with Ireland exhibiting one of the largest relative contributions of population increase to labour force growth. For females, both population and participation components are positive in all countries; only in Ireland and France did the relative contribution of the population effect outstrip the participation rate effect.

It is clear that alongside changes in the size and structure of the population of working age, participation rates play a crucial role in labour force size and structure. One of the key features of socio-economic change over the last twenty-five years has been an increase in participation rates for women (Joshi 1989, Green 1993, Hakim 1993), whereas those for men have remained stable or declined (Meulders et al. 1992; Fig. 4.2). However, participation rates in western Europe remain relatively low compared with those in other industrialized parts of the world. In 1969/70 in the European Community and the USA approximately 65 per cent of people aged 16–65 years were in work; over the following twenty years this proportion rose in the USA, but not in the European Community. The higher proportion of working age population actually in work goes half way towards explaining higher living standards per head of population in the USA compared with western Europe (Sysdem 1990).

Analysts concerned with estimating the future size and composition of the labour force are faced with a greater range of uncertainty in forecasting participation rates than in projecting demographic change. With regard to population projections the main uncertainties involve birth rates and migration patterns; whereas forecasting of participation rates involves forecasting the future level of economic activity, the structure of the economy and the influence of technological change on employment.

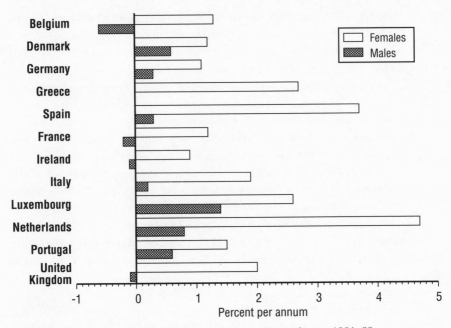

Figure 4.2 Labour force participation mean annual growth rate, 1981–89.
Source: OECD (1990).

In an analysis of demography and the labour force published in 1988 based on
Eurostat data banks, labour force projections were calculated by applying 1985 participation rates to population projections. The projections showed a slight rise in
the total labour force of western Europe up to 1995, followed by a period of stability from 1995 to 2010, and thereafter a steady decline to the end of the analysis
period in 2025. The projected decline in the labour force was most pronounced in
countries such as Germany, Belgium and Italy, and much less important in France
and the UK. Spain and Greece showed almost no change in the total labour force
during the 1990s, while in Ireland there was a steady increase in the labour force
over the entire period. Hence, a decline in the labour force in some countries can-
cels out much of the gain in others. Most of the decline was projected to occur in
the under 25 and 25–44 years age groups. The numbers in the 45–64 years age
group was projected to rise up to 2015, but to decline after 2020.

Labour demand trends

The western European economy has undergone major structural transformation
over recent years. Future economic competitiveness will be determined by the abil-

ity to confront major global challenges – including those from the Pacific Rim and the newly industrializing countries. Combinations of such structural factors and global developments present four main challenges to western European industry. First, standards of living and levels of employment will continue to depend on capacity to stay abreast of international industrial competition. Secondly, firms' capacity to invest more and more efficiently in equipment and technological know-how will continue to be a prerequisite. The third challenge is related to the second: the capacity to master efficiently the diffusion of technological innovation will offer a crucial competitive advantage. From a population perspective, however, it is the fourth challenge that appears (at least at face value) most crucial: the capacity to develop human resources to master technological change and new work organization. So what are the industries of the future, and what are their skill needs?

The latter part of the 1980s was a time of economic optimism for Europe. Between 1985 and 1990, employment in western Europe increased by an average of 1.5 per cent per year – a net addition of nine million jobs (CEC 1991b). All countries recorded positive employment growth over the period, but rates of growth varied considerably – with Spain, Luxembourg, the United Kingdom and the Netherlands recording the highest annual growth rates. However, since the beginning of the 1990s economic growth has slowed appreciably in many areas as Europe plunged into the second recession in ten years. Despite these setbacks, a positive benefit to employment has been predicted from the processes of greater integration inherent in the Single European Market (Cecchini 1988).

Key features of the changing sectoral composition of employment in western Europe are the demise of jobs in agriculture and manufacturing and the growth of services. The proportion of employment in services increased from 42 per cent in 1965 to 62 per cent in 1989, whereas the proportion in agriculture and industry fell from 41 per cent to 32 per cent over the same period. In general, the peripheral regions – exemplified by Greece, Portugal and Ireland – have a higher proportion of people employed in agriculture, and a lower proportion employed in industry than the core regions. In the more peripheral regions, employment in industry and services has to grow at a higher rate than in other areas, not just to cater for the significantly higher growth of working age population and the larger numbers of unemployed (referred to in the next section), but also to compensate for the steady contraction in the number of jobs in agriculture. This problem is intensified by the fact that the proportion of employment in services is particularly high in large cities. In absolute terms, the majority of new jobs in the next ten years are likely to be in business, other private and public services. Household-based services – such as cleaning and childcare – are also expected to increase as more women go out to work.

Alongside the sectoral changes outlined above, during the 1980s there was a trend towards a much more varied mix of working practices, as the number of temporary, seasonal and casual jobs increased. In the United Kingdom, for example, part-time employment grew faster than full-time employment (although this trend has not been common throughout the whole of western Europe). The majority of part-time (and other atypical) workers are women, and the growth in precarious employment and the increasing participation of women in the labour force (examined in more detail in the next section) are interlinked; in 1988 part-time employment represented 13 per cent of total Community employment, but 28 per cent of women's employment. In some countries (such as Ireland, Luxembourg, Italy) rates of part-time working are fairly consistent across the age distribution; in others (such as the United Kingdom, Germany and France) rates of part-time working vary markedly by age group. Differences in part-time working are a reflection of a mix of economic, legislative and social factors. However, there has been a tendency for part-time working to increase particularly rapidly in the most economically successful and prosperous parts of Germany, France and northern Italy – such as Bavaria, the Paris region, and Emilia-Romagna. Part-time work is a contentious subject: many women want part-time work – particularly when their children are young – but some are employed in part-time jobs because no full-time work is available (Conroy Jackson 1991). Therefore, part-time work may disguise underemployment or part-time unemployment. It is estimated that there are 14 million part-time workers and 10 million temporary workers in western Europe.

There is a need for legitimization of such atypical work if there are not to be "Two Europes": one of full-time workers, and another of marginal part-time and atypical workers; particularly as atypical and self-employment are projected to grow relative to full-time employees in employment. These developments are associated with demands for increasing flexibility of the workforce – as epitomized by the emergence of core and peripheral workers. The increasing fluidity of the labour market raises questions about the link between pensions and employment contracts, and there is a clear need to reconcile greater flexibility in the labour market with the need to maintain acceptable levels of social support (an issue highlighted in more detail in the concluding section).

Changes in the sectoral structure have important implications for patterns of change in the occupational structure of employment. However, a further factor is also at work in explaining occupational change: the effect of organizational and technological changes within industries. In recent years these two factors have tended to reinforce one another, leading to a growth in demand for higher level, non-manual skills and a reduction in demand for lower level manual ones. In the United Kingdom a 1.3 million (approximately 15%) increase in employment in the higher level non-manual occupational categories is forecast between 1990 and 2000. By 2000 professional, associate professional and managerial & administrative occupations are expected to account for over one-third of total employment, compared with under one-quarter in 1971 (IER 1991, 1993). Over the medium-term plant & machine operatives are forecast to witness the largest relative and absolute declines in employment, followed by craft & skilled manual workers and other occupations; such that by 2000 it is estimated that these three occupational categories will account for under one-third of total employment, compared with nearly one-half in 1971. Similar broad features of occupational change are likely to be prevalent in other parts of western Europe.

Both within individual countries and at the pan-European scale a complex array of factors will influence and change the spatial distribution of the demand for labour. At the western European scale, two main core areas likely to continue to attract new firms are identifiable: first, a triangle of Paris, London and Amsterdam (including the Ruhr valley); and secondly, southern Germany, northern Italy, parts of southern France and areas around Barcelona and Valencia (CEC 1991b). Alongside such pressures for centralization, there are also important forces for decentralization, which contain the potential for a more even distribution of activity (Masser et al. 1992). These include the spread of flexible production systems, a forecast growth in the number of small firms, rising costs of congestion in core areas, the declining numbers in prime working age groups in certain core areas, whereas new developments in advanced forms of transport, telecommunications and integration of energy transmission networks are likely to help overcome the disadvantages of some of the more remote regions.

Looking ahead to the early years of the twenty-first century, the key question is whether pressures arising from economic and geographical imbalances between prosperous centres and the rest of the Community will encourage the mobility of economic activity and jobs and/or the mobility of labour, and to what extent barriers will be imposed to restrict such movements. The enhanced locational freedom implied in a *"Europe sans frontières"* suggests that changes may take place in the relative prosperity of regions and local economies on a European scale. On the one hand, by stimulating the European economy the Single European Market and associated processes of integration have a vital role to play in combating regional underdevelopment, whereas on the other, the increased priority and resources allocated to European structural funds is an acknowledgement of the possibility of aggravated regional imbalance, with many regions adversely affected by intensified competitive pressures following on from the completion of the Single European Market. The consequences for labour demand and supply subgroups at the local scale will be a function of changes in competitiveness at a variety of spatial scales: the competitiveness of Europe in the wider world, changing interregional and inter-urban fortunes at the national and international scales, the changing balance of advantage between core and peripheral areas, and between urban and rural areas. Expanding employment sectors will not necessarily locate in the same areas as those undergoing employment decline – creating severe short-run (and perhaps long-run) crises of unemployment (Green 1992).

Labour supply and demand interactions

The best known and most commonly used indicator of mismatches in the interaction of labour supply and demand is the unemployment rate. In April 1991 the unemployment rate in western Europe was 8.8 per cent (12 million people) – compared with 8.4 per cent in April 1990. Increases in the year 1990–91 were particularly pronounced in the southern regions of the United Kingdom, and in parts of Spain and Italy (Eurostat 1991b). Although the regional rates had converged during the recovery in the second part of the 1980s, there remained considerable regional disparities and unemployment rates still exceeded 15 per cent in parts of Spain (mostly the coastal regions), southern Italy, the Republic of Ireland and Northern Ireland (Fig. 4.3); (in Ireland the rise in unemployment was reduced throughout the 1980s by relatively high rates of out-migration of young people (often with higher-level qualifications) to the United Kingdom, western Europe and the USA). Higher rates of unemployment in the less developed regions are in part related to demographic trends: higher birth rates in these regions continue to result in faster growth in the labour force than elsewhere. Hence, stronger employment growth is needed in regions lagging behind to offset the relatively faster growth of the labour force before unemployment disparities with the rest of Europe can be reduced.

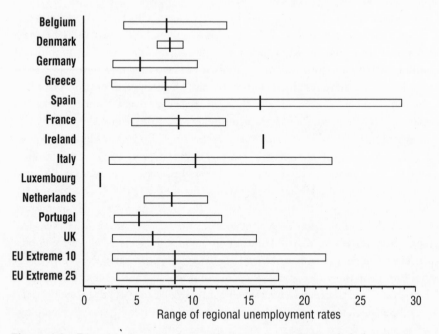

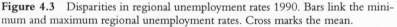

Figure 4.3 Disparities in regional unemployment rates 1990. Bars link the minimum and maximum regional unemployment rates. Cross marks the mean.

Alongside the uneven geographical distribution of unemployment (Ballesteros 1993, Vandermotten 1993), a differential incidence of unemployment by subgroup is evident. Except in the United Kingdom, unemployment rates in western Europe tend to be higher for women than for men. Overall, more than half the unemployed in Europe are women. There appears to be a relationship between the incidence of unemployment and the number of part-time workers – it could be that part-time employment helps to "mop up" female unemployment in countries such as the United Kingdom. Youth unemployment is most severe in parts of Spain and Italy, and remains a significant problem in Portugal, Greece and Ireland. In Germany, Belgium, the Netherlands, Denmark and the United Kingdom, long duration unemployment is largely a problem of adults – stemming from problems of industrial restructuring and the re-integration of older workers. Overall, long-term unemployment has contributed to serious problems of poverty among middle and younger age groups in recent years.

The effect on unemployment of differential rates of employment growth across Europe has not been uniform. Only 30 per cent of the new jobs created in western Europe between 1985 and 1990 were taken by those recorded as previously unemployed – hence unemployment rates have remained stubbornly high, and (as indicated above) long-term unemployment represents a continuing problem. The somewhat weak and delayed response of unemployment to the recovery in output in the 1980s was primarily a reflection of the upward pressure of demographic factors on labour supply, together with the general rise in economic activity rates.

Unemployment is considered by some commentators a relatively narrow measure of labour market mismatch; unemployment statistics do not include those who might want a job should suitable jobs become available or those on government (or similar) training schemes. A broader measure of the unutilized labour force is the ratio of employment to working age population. In places where this ratio is low, it is reasonable to assume that there are relatively large numbers of economically inactive people who would like to work if employment opportunities were created (whether or not these people are included in the unemployment statistics). In 1989 the proportion of working age population in employment was 62 per cent in the "more developed" parts of the European Community, compared with 50 per cent in the "less developed regions" (those classified as Objective 1 for Structural Policy purposes). The ratio of employment to working age population rose on average by three percentage points in the "more developed parts" of the Community between 1985 and 1989 and by only 1.5 per cent in the "less developed regions" – indicating a divergence in the relative sizes of the unutilized labour force in the two types of area. On a similar theme, the Commission of the European Communities has made estimates of hidden labour supply/underemployment based on the conservative assumption that a hard core of 20 per cent of the working population would remain inactive irrespective of the number of jobs on offer, and irrespective of the measures taken to facilitate participation. Estimates suggest that hidden labour supply was significant in almost all areas outside Denmark in 1989 – ranging from over 10 per cent in much of France and Germany, 15 per cent in northern Italy, and over 20 per cent in Spain, southern Italy and Belgium. Taking hidden labour supply and unemployment together, just under 30 per cent of people of working age were actual or potential job seekers in the "less developed regions" in 1989, compared with under 20 per cent in other areas.

Unemployment results when labour supply outstrips demand. Another factor that can be an important factor in labour supply and demand interactions is migration. Migration patterns are determined by a complex of economic and social factors, including differences in employment opportunities and wage levels, and policy stances towards migrants (European Foundation for the Improvement of Living and Working Conditions 1990, King 1993b, Simon 1993). Labour mobility from areas with an excess supply of labour to areas with an excess of demand is one possible approach to labour market imbalance. Just under 5 per cent of the total population of the European Community at the start of the 1990s was of a nationality other than that of the host member state, of whom about one-third were from other member states. Up to this time, the main "supplying" countries within the Community were Ireland and Portugal. Despite the elimination of restrictions to the movement of labour between countries of the European Union – with the avowed aim of widening employment opportunities and ensuring a more balanced labour market – labour movements have never been seen as a major means of correcting labour market imbalance.

Nevertheless, a number of particular issues and developments are identifiable as worthy of note for present and future migration trends. First, there is the very large potential for immigration from eastern and central Europe to western Europe. The working age population potentially available to work in eastern and central Europe is around 90 million (about 40% that of the European Community at the end of the 1980s), and in these countries prior to democratization and break-up economic activity rates were very high. A further major uncertainty over the medium-term concerns the potential pressure for legal and illegal migration into western Europe for political and economic reasons from more or less developed and newly developing countries in Latin America, Africa and Asia. Furthermore, there is a secular trend towards higher mobility of the world population caused by a reduction in travel costs and improved information about foreign countries. How many potential immigrants will take serious steps to move will be determined largely by factors such as the widening of economic distance, the reduction of cultural distance, and the force of legal barriers. The European Union countries are adopting an increasingly "schizophrenic" stance: characterized by the open borders of the "Schengen group" and authoritarian actions to keep out asylum-seekers and economic migrants, co-ordinated at government level. Whether immigrants can be integrated by Community countries will be determined largely by the demographic aspect of future population development – immigration may help compensate for a deteriorating age structure (immigrants generally have a younger age structure than the host population – with consequent implications for labour force growth), the economic effects of a growing population and labour force, and the qualitative structure of the immigrating labour force. The current trend is for labour flows of more highly qualified people (many making intra-organizational moves). On balance, it would seem that the underlying pattern of slow but persistent inward migration seems set to continue in the future.

Key labour market issues and subgroups

ISSUES

One of the foremost issues at the intersection of demographic and labour market change at the turn of the twenty-first century is the decline in the younger working age groups and the consequent ageing of the labour force. As the economy expanded once again in the second half of the 1980s, the term "the demographic time bomb" was coined to capture the effect of the downturn in the number of young people entering the labour market in the 1990s and beyond (National Economic Development Office/Training Commission 1988, National Economic Development Office/Training Agency 1989, Green & Owen 1993). Both demographic factors and participation rates have an influence on net entrants to the labour force. Holding the effect of participation rates constant and so considering the effect of demographic factors alone, the excess of new entrants to the labour force (mainly young people) over leavers in western European peaked at 1.21 million in 1981. On the same assumptions, the projections are for a reversal of this situation, with a deficit of almost the same level by 2025. With the downturn in the number of young entrants to the labour market, it is necessary for employers to consider other options, such as recruiting older workers and encouraging women returners to enter the labour market (these subgroups are discussed in more detail below), and re-integrating the long-term unemployed. Labour market analysts considered that the demographic time bomb could herald two main trends (or a mixture of both): first, it could be a constraint on growth – with skills/labour shortages ensuing; or secondly, it could be a major opportunity to take a more positive approach to expanding employment and incomes, and reducing unemployment, particularly if appropriate training were to be made available. As the European economy entered recession once again at the beginning of the 1990s, the spectre of the demographic time bomb became less apparent in the context of overall job loss

63

and the downturn in the number of young people on the job market moved down the policy agenda.

Although they are not a new phenomenon, the implications of demographic trends on labour supply and the threat that this poses to Europe's competitive strength, at a time when other major industrial and technical changes are influencing employment and the labour market, has led to growing concern about skill shortages. However, skill shortages are difficult to define and even harder to measure. Business surveys suggest that skill shortages have become more acute since the mid-1980s (a time of economic expansion and tightening labour markets) in most parts of western Europe. From a policy perspective, the way in which short-term problems of labour shortages are tackled will have implications for longer-term problems and their solutions.

It is easy to focus on education and training as a panacea for skill shortages. A high level of educational attainment represents the foundation for the necessary level of human capital that advanced economies require in the face of the occupational changes outlined above. It is clear, however, that life-long learning is of increasing importance. There are examples of imbalances in supply and demand of trained people in all parts of Europe – even in economic downturns, and adult training could be a key issue in determining whether older workers and women obtain a fair share of the future benefits of economic growth. Research has shown that the employees most likely to be in receipt of training are the young, and those in professional occupations; in taking a break from paid work many women "miss out" on some or all training. Current imbalances in the quantity of training across western Europe are considerable. The proportion of people receiving training in 1989 varied from under 2 per cent in Greece to over 15 per cent in the Netherlands and Denmark. Moreover, the quality of training offered is important. To bring education and training infrastructures in lagging regions up to average standards (both quantitatively and qualitatively), a major regionally differentiated investment effort is needed.

SUBGROUPS

In the face of rising real incomes and earnings, and because of continuing social trends towards a greater labour market role for women (see below), changes in economic activity rates for older workers are expected to result in reductions in the supply of older men (by approximately 250000 in the United Kingdom), but increases in older women (of approximately 150000 in the United Kingdom) by the turn of the century. Fortunes of older workers in the labour market will depend crucially on the qualifications and experience they possess. The increase in self-employment may favour older workers: due to the stocks of contacts, experience and skills many of them have built up throughout their working lives. In the United

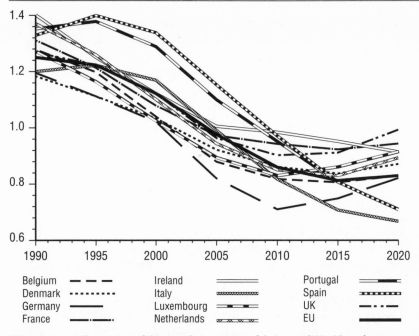

Belgium	— — —	Ireland	═══════	Portugal	══════
Denmark	··········	Italy	▨▨▨▨▨▨	Spain	■■■■■■■
Germany	————	Luxembourg	═ ■ ═ ■	UK	— · ─ · —
France	—·—·—·	Netherlands	⨯⨯⨯⨯⨯	EU	▬▬▬▬▬

Figure 4.4   The ageing of the population: ratio of those aged 20–39 to those aged 40–59. *Data source: Eurostat* (1991c).

Kingdom the net age demand effect (i.e. the change in employment by age due to changing age proportions within each industry and occupation over and above those changes that would have taken place if the age structure had changed in line with the underlying aggregate demographic and activity rate trends) was positive for older workers, but strongly negative for younger people, between 1979 and 1983 – indicating that reduced recruitment of younger people (partly because employers find such workers are relatively expensive) typically outweighed the impact of quits and redundancy for older workers (Lindley et al. 1991). During the subsequent recovery (1983–8) this process was reversed.

With regard to consequences of an increasing proportion of older workers (as highlighted in Figure 4.4), there are fears that the ageing process will involve greater conservatism in business, that processes of innovation and adaptation will be hindered, and that wage costs will be forced to rise and productivity fall. Two main aspects deserve particular consideration. First, there is the impact of ageing on skills gained through professional qualifications and working practice: at a time of rapid technological change, imposing the need for rapid updating of skills, older people are in danger of finding themselves at a disadvantage compared with young people (despite the fact that research has shown that acquired experience and competence are not rendered completely obsolete by technological change). Secondly, there is the impact of ageing on economic performance of a company: which is in danger of being increasingly negative as wage costs rise in line with age, whereas productivity falls. In this respect there is a case for a revision of certain traditional principles concerning the management of careers and the fixing of salaries.

Alongside the growth in older workers, another group of increasing significance in the workforce is women. Feminization has been a key feature of the European workforce over the last thirty years. By the late 1980s women accounted for 45 per cent of the workforce in western Europe (however, the employment rate for women lagged 15 percentage points behind that in eastern Europe). The feminization of the workforce is mainly due to increasing numbers of married women in paid work. As the demographic downturn results in a smaller cohort entering the workforce during the 1990s, it is projected that the greatest single element of growth will be women returners – those returning to paid work after a break for childbirth and child-rearing. Of the 4.8 million extra jobs created in Europe in the late 1980s, six out of ten went to women. Women's employment increased at twice the rate for men between 1985 and 1990, although in southern and eastern Spain, Brittany and Northern Ireland where employment growth was relatively high, less than half of the additional jobs went to women (CEC 1992).

Many barriers remain preventing women having the same access to jobs as men. For women to work, social support, training, and available jobs, are needed. Moreover, the quality of work open to women is a key issue. The European Union has been developing policies for women since the 1970s. These were based on Article 119 of the Treaty of Rome, calling for equal pay for equal work, and subsequent Action Programmes on Equal Opportunities have been introduced in recognition of the need to tackle the specific issues facing women in the labour market. Childcare is also an important issue. Women are tending to have fewer children, to have those children later in life, and to want a better life for those children they do have. They are also tending to return to work sooner after, and between, births of children. As regards maternity leave, there is a need for a double strategy of embracing more women in the labour force, but allowing greater scope to allow them to exit the labour force as and when they desire. Since there are likely to be shortages of skilled workers in some regions that can only be filled by women, childcare will be vital for women's access to new occupations.

Selected policy issues

Ageing of the population appears to present a strategic problem (as outlined above), and this is why it has come to be of major scientific and policy concern. In the recessionary conditions of the early 1980s, older workers were perceived as victims of the labour market – as they were affected by redundancy and early retirement, whereas the emphasis has begun to shift such that older people are perceived as a potential burden on the wider economy. Questions relating to older people (and other subgroups) need to be considered more in the context of the individual's pattern of opportunity and experience over the whole life-cycle rather than to treat specific sociodemographic periods and subgroups in isolation.

Despite the tendency to deal with catch-all categories such as "older workers" and "women", policy questions are as much about dealing with differences among subgroups, as dealing with problems they share in common. There is evidence for growing polarization within and between groups and areas. For example, there are wide differences in the choices available for older people and women with good career jobs and those who have poorer conditions of employment or who are unemployed.

Unemployment problems need to be addressed by policies modulated according to a wide variety of regional circumstances. In areas such as southern Italy, Spain and Ireland, the problems of an ageing population and labour force are less pressing than the problems of high rates of youth unemployment and the relatively greater flow of young people onto the labour market. Providing training and qualifications for this group is a priority. For many northern regions, declining numbers of young people and associated ageing of the workforce requires an emphasis on the creation of facilities for continuous education during adult life, more targeted opportunities for training, and the breakdown of discriminatory labour market practices, since these countries also have growing youthful minority ethnic group populations, often with some of the highest rates of qualifications – for example, Black-Africans in Great Britain have amongst the highest share of highly qualified people of any ethnic group, but also suffer higher than average unemployment rates (Owen 1994).

Conclusions

This chapter has emphasized that demographic change is only one element in the labour market. Relative to changes in participation rates, the impact of demographic change is likely to be of diminishing significance as Europe approaches the turn of the century. Participation rates tend to fluctuate in response to economic conditions, but it is likely that those for older workers will tend to decline overall, as demand for the labour of older people with lower productivity and specific skills contracts. However, participation rates of younger people may not grow greatly, as a result of initiatives to increase training and educational standards (which also delay the entry of young people into the labour market). Participation rates of married women are likely to continue to rise, due to falling family sizes and the need to provide or augment household incomes. Further contraction of the agricultural sector in Europe, especially in less developed areas, may raise participation rates as people currently working in an informal fashion on family farms are forced into the formal labour market.

The demographic time bomb/shock/crisis has acquired a slant and status in policy circles and the media that it does not deserve (Haughton 1990). The fall in the number of young people who are potential labour market entrants is often measured from the top of the demographic peak to the bottom of the trough, thus exaggerating its significance. It is being seen as a potential problem to which labour market, social security and training policies should be addressed. However, it would seem more appropriate to regard certain elements of these policies as themselves turning demographic change into a potential crisis: the distortions produced by old policy measures may be a cause of a problem, rather than new policy measures a solution (e.g. in the case of pensions). The current social security systems are designed for a historic type of labour market – in which men worked full-time in steady employment, and women gave up work when they had children. This is now out-of-step with reality.

The redistribution of employment across industries, occupations, geographical areas and people with different socio-economic characteristics has the potential to worsen the imbalance of employment opportunities, and associated inequalities, that already exist within and between subgroups and areas. As the economic relevance of national boundaries progressively diminishes, the regional balance in the demand for, and supply of, labour will increasingly become a matter of general policy concern. Large movements of labour from one area to another are liable both to widen disparities in economic performance, as the workforce in weaker regions is depleted, and to add to congestion and environmental problems in stronger regions. Harmonization of policies is likely to have as a consequence higher standards, aggravating problems of unemployment in countries and regions with less developed economies and intensifying polarization between privileged and disadvantaged regions and groups of people. There is a need for massive education and training programmes specially designed for disadvantaged groups and areas that will enable them to take advantage of the challenges of the new era.

From a political and citizenship perspective, there is a problem in reconciling the rights of migrants (both internal and international) with those of indigenous workers and those in areas added to a growing European Union. These tensions will be most severe in the more economically successful parts of the European Union, and if economic growth is not maintained in these regions, pressure for more restrictive migration policies and nationality laws will grow.

CHAPTER 5

Spatial inequalities of mortality in the European Union

DANIEL NOIN

Introduction

Research on the spatial inequalities of mortality within Europe is still scarce. There are a few partially detailed studies on some countries, but almost none at the level of the countries of the European Union as a whole. Geographical aspects of mortality have hardly been considered so far, and the rare existing information has been based on national figures showing relatively small differences between countries, rather than on regional data within countries.

More detailed analysis would be useful. How wide are spatial inequalities? How are they evolving, and what are the main trends? What are the prospects for the year 2000? It is to these questions that this chapter seeks an answer, using a variety of statistical information currently available, including the United Nations Demographic Yearbooks and the volumes on World Population Prospects produced in 1988 and 1990, alongside the World Population Data Sheets produced by the Population Reference Bureau in Washington and, for Europe, the Demographic Statistics prepared by Eurostat in Luxembourg.

An assessment of the current situation

The level of mortality and life expectancy in the European Union as a whole can be considered as satisfactory in absolute terms. However, when comparison is made with other countries of the developed world (Table 5.1), Japan is undeniably the most advanced in both infant mortality and life expectancy. The EU results are good but are not among the best. No European Union country can be seen in the top ranks for life expectancy and several countries are far down the list. Although the poor ranking of Portugal and the former East Germany is not too surprising (31st and 28th places respectively on the world ranking for the end of the 1980s), that of

Table 5.1 Mortality indicators for the developed world.

	Infant mortality (per thousand live births)		Life expectancy at birth (in years)	
	1980–85	1988–9	1980–85	1988–9
Japan	7	5	76.5	79.0
Scandinavia (excl. Denmark)	8	7	75.7	76.7
Australia/New Zealand	10	8	74.9	76.0
European Union	11	8	74.3	75.8
USA/Canada	11	9	74.5	75.5

Sources: United Nations, *Demographic yearbooks.*

Table 5.2 Mortality indicators for the countries of the European Union.

	Infant mortality (per thousand live births)		Life expectancy at birth (in years	
	1980–85	1988–9	1980–85	1988–9
Ireland	10.0	7.5	73.1	74.6
United Kingdom	10.7	8.4	74.0	75.6
Denmark	8.2*	6.0*	74.5	75.8
W. Germany	10.8	7.5	73.9	75.2
E. Germany	11.0	8.1	72.1**	73.7**
Netherlands	8.3*	6.8	76.0*	77.1*
Belgium	11.0	8.6	73.7	75.1
Luxembourg	10.3	7.0	73.3	74.7
France	9.2	7.2	74.7	76.1
Portugal	21.1**	12.1**	72.2**	73.8**
Spain	10.6	8.3	75.8*	76.9*
Italy	12.6	8.8	74.6	75.9
Greece	16.0**	9.8**	74.7	76.0

Notes:
* good performance – one standard deviation below infant mortality average for EU, one standard deviation above life expectancy average for the EU.
** poor performance – reverse of the above.
Sources: Eurostat, *Demographic statistics.*

the UK and West Germany is more so (in 22nd and 20th place respectively). These are poor performances, considering the large financial investment in health, social protection and education systems that have been made in Europe: questions must obviously be raised about other issues affecting mortality, such as diet.

The overall figures for the EU shown in Table 5.1 clearly mask quite distinctive spatial inequalities at the national level (Table 5.2). The differences are greater in the figures for infant mortality. In the period 1980–85, the worst performing coun-

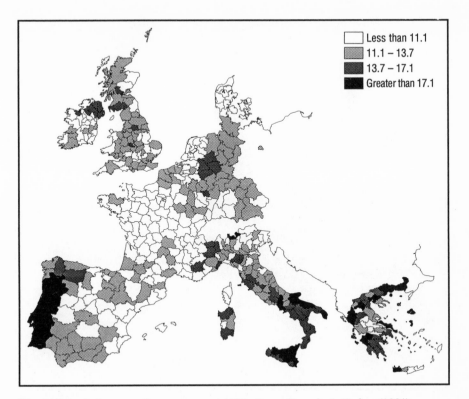

Figure 5.1 Infant mortality rates around 1980. *Source:* Decroly & Vanlaer (1991).

try, Portugal, had an infant mortality rate two and a half times that of the best, Denmark. These disparities are much greater when subnational divisions such as English counties or French *départements* are taken into account (Decroly & Vanlaer 1991), as Figure 5.1 demonstrates. The advanced status of the Netherlands, Denmark and France is easily seen on the map with uniformly light tones, whereas Portugal is clearly lagging behind: here the tones are uniformly dark. In other countries the situation is more contrasted, with certain regions of low infant mortality and others at much higher levels, for example in Germany, Spain, Italy, Greece and the UK. At this subnational scale the ratio of the highest to lowest infant mortality rates is 5:1, between some Danish districts around Copenhagen and certain parts of northern Portugal.

Life expectancy at birth differences are not very marked when national results are considered. The maximum difference between countries in 1980 was 4.4 years, reducing to 3.4 years by the end of the decade. Ignoring East Germany, which only became part of the EU on German reunification in October 1990, the Netherlands and Portugal again occupied the first and last places on the list.

71

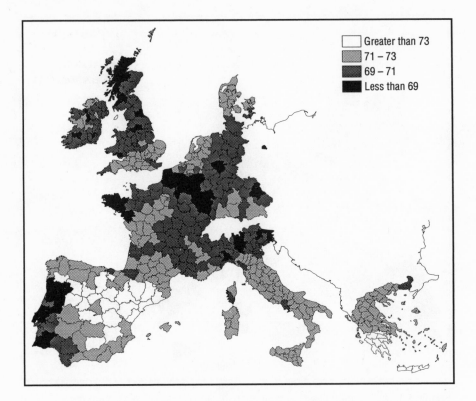

Figure 5.2 Male life expectancy at birth around 1980. *Source:* Decroly & Vanlaer (1991).

However, these inequalities are naturally more pronounced if finer divisions are considered, with life expectancy at birth estimated from standardized mortality indices calculated by the indirect method (Poulain 1990) (see Figs 5.2 and 5.3). For women in 1980, life expectancy was already over 80 years in some geographical units, especially in the northern half of Spain, whereas it was less than 75 in most of Ireland, and in some parts of Portugal and the UK. For men, at the same date, the equivalent high and low life expectancy ages were over 73 and under 68 years. The effect of national policies is quite evident: some countries are markedly distinct from their neighbours, with political boundaries showing up significantly. Portugal attracts attention by its poor figures, whereas Denmark, the Netherlands, France and Spain display generally good results. Internal inequalities are clearly marked in the United Kingdom where the South East is contrasted with the rest of the country, and in Belgium the classic opposition between Flanders and Wallonia is present (Decroly & Grimmeau 1991). More surprisingly, in Italy the disparity between the north and the rest of the country is the reverse of what might be expected, with the southern part having better figures.

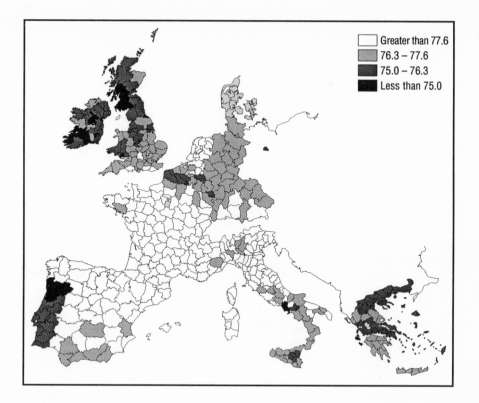

Figure 5.3 Female life expectancy at birth around 1980. *Source:* Decroly & Vanlaer (1991).

The factors explaining differences in life expectancy are many and complex, and no full study giving an overview of the situation in Europe has so far been completed (van Poppel 1979, Bouvier-Colle & Jougla 1989). The relationship between life expectancy and average wealth level is not strong: Luxembourg has the highest gross national product per capita in the EU but ranks only tenth for life expectancy, whereas Spain is in second position, with a per capita gross national product only just over one-third as high. The link with the quality of health and social protection is weak too, as shown in certain studies (Vallin 1984, Kunst et al. 1988): several Mediterranean countries, where such systems are poorer than in northern Europe, are nevertheless classified high up in the life expectancy figures. The relationship with the average level of education does not seem particularly strong: in this respect also, the results for the Mediterranean countries are puzzling.

In fact, the essential explanatory factors are now generally believed to be linked to behaviour and habits, in particular to smoking and the consumption of alcohol and animal fats, which are generally greater in the northern regions of the EU than in the southern. Several examples can be used to illustrate this point. First, death by cancer of the digestive system is undeniably linked to alcohol consumption, and is six times greater in the calvados drinking regions of northwestern France than in southern Italy. Secondly, death from lung cancer is highly significant in the UK, Benelux and Germany where cigarette smoking was widespread and started early among youngsters at least a generation ago. However, with falling levels of cigarette consumption in northern Europe and increased levels in the south it is likely that this mortality differential will narrow in the future as lung cancer deaths in the south grow in importance.

Thirdly, death from cardiovascular diseases is much more significant in countries with a high consumption of animal fats, for example in the British Isles, Belgium, Germany and northern France, than in those countries where food is generally cooked in olive oil, as in the Mediterranean regions. The improvement effect of the Mediterranean diet appears to be more important for men than for women: for example, in parts of southern Italy (eastern Sicily) and Greece (Macedonia and the Aegean Islands) male longevity is better than the EU average for all men, while that of females is below the average for all women. This depresses sex differences in mortality in these regions (see below).

Fourthly, it has also been observed that death from cancer and cardiovascular diseases is weaker in southern regions where fruit and vegetable consumption is traditionally higher (Broard & Lopez 1985).

This analysis of factors nevertheless requires more work through further research. One aspect of this research should be the preparation of detailed maps of mortality by age, sex and cause of death. Some of these maps already exist (Holland 1988) but more extensive research is needed, with a potential for the use of geographic information systems for the suggestion of hypotheses on causal interrelationships.

The inequality of mortality between the sexes is a further very marked feature within the European Union. In the early 1980s the difference in life expectancy between men and women stood at 6.7 years in the EU, against 5.5 years in Japan, 6.1 years in Scandinavia (excluding Denmark) and 6.6 years in Australia and New Zealand. Of the regions of the developed world only the USA and Canada had a higher sex difference at 7.4 years.

Male mortality is higher than female at all ages in the EU, but especially for young people and the elderly. For the young the difference is particularly explained by accidents and violent deaths. For the elderly, explanations are more likely to be cancer and cardiovascular diseases, themselves linked to differences in behaviour, especially with regard to tobacco, alcohol and diet.

Several authors attempt to explain the differences through smoking habits (Fox 1978). In fact, the causes are more varied. Smoking no doubt accounts for several years of difference in life expectancy between men and women in the northern part of the EU, where it used to be more widespread twenty or thirty years ago and

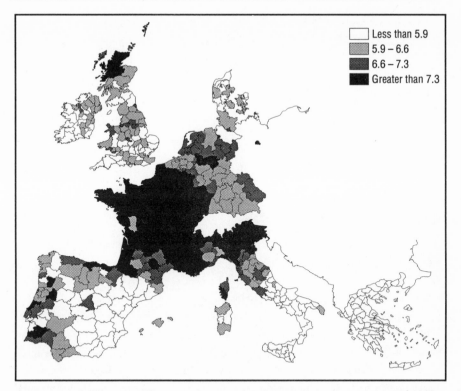

Figure 5.4 Difference in life expectancy at birth between men and women around 1980. *Source:* Decroly & Vanlaer (1991).

where people started smoking earlier. But alcohol use, and abuse, is undoubtedly responsible for some of the differences, considering the marked disparities we can observe in certain regions of Europe such as northwestern France, northern Italy and the northern fringe of Spain (Fig. 5.4).

Variations at the national level are, in fact, quite important (Table 5.3). Greece's figure is the lowest at one standard deviation below the mean for the EU; that for France is the highest, significantly above the mean. Indeed, France has the highest sex difference in longevity in Europe outside the former Soviet Union (where the difference in the early 1980s was 9.2 years). The next highest difference was in Poland at 8.0 years.

As Figure 5.4 shows, at subnational scales these variations are even more pronounced. The greatest differences in life expectancy are observed in certain French and Italian districts where the gap is almost ten years in some places (Finistère and Morbihan in France, Belluno in Italy). This gap is less than five years in some other regions in Italy and in Greece (Sicily and the Italian South; the Ionian Islands, the Peloponnese and Crete). Low differences also existed in East Germany, in parts of southern Spain, and in parts of Ireland.

75

Table 5.3 Differences of life expectancy between men and women (in years), 1980–85.

Ireland	5.5
United Kingdom	6.2
Denmark	5.9
W. Germany	6.7
E. Germany	5.9
Netherlands	6.7
Belgium	6.8
Luxembourg	6.7
France	8.2
Portugal	7.0
Spain	6.1
Italy	6.6
Greece	4.2

Sources: Eurostat, *Demographic statistics* (1990).

National studies suggest that these sex differences in relation to mortality are resistant features which are not disappearing. Amongst men, the socioprofessional groups that are most favoured for longevity are executives and those in liberal and intellectual professions, whereas the least favoured are generally unskilled manual workers. Social inequalities of mortality seem to be linked more to the level of education and training and to life-styles, than to income levels (Desplanques 1984, Valkonen 1987). Less is known about the causes of variations in female longevity.

Evolutionary trends

Before looking at the prospects for mortality and life expectancy in the year 2000, we should first make some brief comments on the trends and factors of evolution.

There has been a considerable drop in infant mortality over the post-war decades throughout western Europe, at about 5 per cent on average per year. Its level is now only a quarter of that in 1950 (Table 5.4). Life expectancy at birth has also seen considerable changes. It has been lengthened by nine years on average for western Europe over the same period, with an increase of just under three months of life every year. This progress has not, however, been regular: it was rapid in the 1950s and early 1960s, slower in the late 1960s and early 1970s, and has been quite fast again since the late 1970s (Caselli & Egidi 1981, Casper & Hermann 1991). There

Table 5.4 Evolution of mortality indicators since 1950 for countries of the European Union.

	1950–55	1970–75	1980–85	1988–9
Infant mortality rate (per thousand live births)	49	19	11	8
Percentage annual fall in infant mortality rate since previous period	–	4.6	5.3	5.2
Life expectancy (in years)	66.9	71.7	74.3	75.8
Annual increase in life expectancy (in years) since previous period	–	0.24	0.26	0.25

Sources: United Nations, *Demographic yearbooks.*

Table 5.5 Evolution of life expectancy for the two sexes (in years) for countries of the European Union.

	1950–55	1970–75	1980–85	1988–9
Men	64.7	68.8	71.2	72.6
Women	69.2	74.9	77.8	79.3
Difference	4.5	6.1	6.6	6.7
Average annual increase of life expectancy (in years) since the previous period:				
Men	–	0.21	0.24	0.23
Women	–	0.28	0.29	0.25

Sources: United Nations, *Demographic yearbooks.*

is at present little sign of a slowing down of the rate of mortality improvements. Where they are occurring, falls in the levels of smoking, decreases in the consumption of alcohol and animal fats, together with better levels of education and more frequent practising of physical exercise may be seen as favourable influences for the years to come. These may, however, be partly balanced by increases in mortality attributable to AIDS (see Ch. 6).

Although trends are favourable as a whole, there are quite clear disparities between the two sexes. Since the early 1950s, life expectancy has increased by 10.1 years for women but only by 7.9 years for men (Table 5.5). The difference in longevity between the two sexes has thus significantly increased but with a recent slowing down of this rate of increase. In particular, the rate of increase in female longevity was lower in the 1980s than in earlier post-war years, possibly resulting from the increase in smoking amongst women that occurred in the 1960s. In general, consumption habits (of food, alcohol and tobacco) have moved closer together for the two sexes, and we might expect some convergence in rates of mortality change in the future. This does not mean, however, that the sex differential will necessarily diminish.

After the Second World War mortality differences between countries appeared very marked. Some were in advance for life expectancy, notably the Netherlands and Denmark, while others were far behind, particularly Portugal and Spain. These differences have markedly diminished as the lagging countries have made remarkable progress. Between the early 1950s and the early 1980s the increase in life expectancy was 4.4 years in Denmark, whereas it was 14 years in Portugal (Table 5.6). Hence, a process of convergence is occurring. Amongst the countries of the EU as constituted in 1994, the variation in life expectancy at birth stood at 12.8 years in the early 1950s; a decade later it had fallen to 9.2 years, and by the early 1970s it was 6.0 years. In the early 1980s the variation was 3.9 years and by the end of that decade it was down to 3.4 years.

Table 5.6   Average annual increase in life expectancy (in years) by country.

	1950–55 to 1970–75	1970–75 to 1988–9
Ireland	0.22	0.18
United Kingdom	0.14	0.19
Denmark	0.13**	0.12**
Netherlands	0.10**	0. 19
Belgium	0.20	0.21
Luxembourg	0.23	0.27
W. Germany	0.16	0.29
E. Germany	0.21	0.12**
France	0.30	0.26
Portugal	0.44*	0.42*
Spain	0.45*	0.22
Italy	0.30	0.26
Greece	0.32	0.19

Notes:
* good performance – rate of increase one standard deviation above the average
** poor performance – rate of increase one standard deviation below the average.
Source: Eurostat, *Demographic statistics.*

Thus, the differences between countries have not disappeared completely, but it is likely that they will go on diminishing into the future. Thereby the annual gain in life expectancy has varied quite distinctly from one country to another, with an increase of 0.24 years per year on average between 1950–55 and 1988–9 (Table 5.6).

Prospects for the beginning of the twenty-first century

We can now contemplate the prospects for mortality and longevity at the beginning of the twenty-first century. As always in population projection and forecasting, this is a delicate matter, but for mortality the risk of error is limited because of the relative regularity of its evolution. Certainly the initial development of AIDS mortality appeared to pose a threat to that regular evolution, but as is suggested elsewhere in this book (Ch. 6), the prospects may be less pessimistic than at first feared.

Let us first examine the perspective drawn up by the Population Division of the United Nations for the years 2000 to 2005, for those countries or groups of countries with the most favourable situation regarding mortality (Table 5.7): these perspectives have become one of the general bases for much futures thinking on demographic trends (see also Lutz 1991).

Table 5.7   United Nations demographic predictions for 2000–2005.

	Infant mortality (deaths per thousand live births)	Life expectancy for the 2 sexes (in years)	Difference in life expectancy between the two sexes (in years)
Japan	5	79.6	5.8
Scandinavia (excl. Denmark)	5	78.4	5.8
Australia/New Zealand	6	78.0	6.0
USA/Canada	6	77.7	6.4
European Union	6	77.5	6.1

Source: United Nations, *World population prospects* (1990).

For infant mortality we can justifiably bank on a continuing fall following the steady rhythm observed since the early 1970s, perhaps with a slight slowdown at the end of the century, as the levels will become very small in the most advanced countries. Substantial progress is expected in extending prevention measures, and in raising the level of education of parents. It seems justified to count upon a drop of 4.5 per cent per year, which will result in a rate of 4.5 per thousand live births in the year 2000 – a better level than the UN prospect.

For life expectancy we can also justifiably anticipate substantial improvement at least into the beginning of the twenty-first century. The contribution from the decline of infant mortality will be small as the figures are low already, but we can reckon on a better level of education, on better life-styles, on a general reduction in smoking and in alcohol abuse, and finally on the improvement of geriatric care. It is not unreasonable to foresee the increase in life expectancy following the pace observed over recent years (an increase of one year every four years). The average life expectancy would in this case reach 78.6 years in the year 2000, a figure which is below, but not far below, that of present-day Japan. This figure is about one year

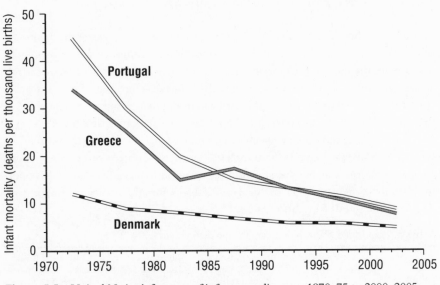

Figure 5.5 United Nation's forecasts of infant mortality rates 1970–75 to 2000–2005 (EU countries with highest and lowest rates only).
Source: United Nations, *World population prospects* (1990).

higher than the United Nations forecast. The European Union would then be about 12 years behind Japan in terms of infant mortality levels, and around 15 years behind in terms of life expectancy.

Overall it is more difficult to foresee the difference in life expectancy between men and women at the beginning of the third millennium than to forecast infant mortality and overall longevity. A change in trend can be expected: the difference between the two sexes could soon reach its maximum and then start to diminish. It is on this hypothesis that the United Nations Population Division has based its outlook. Nevertheless, it is difficult to know, even approximately, when this reversal in trend will occur and at what speed it will be accomplished. If we consider the disparity to be principally attributable to behavioural differences, it would be deceptive to count on rapid change. There is certainly an increase in female consumption of cigarettes and a drop in male consumption of tobacco and alcohol in most European countries, but these modifications are slow. For certain factors such as accidents and suicides, there is no perceptible change. It does not seem justified, therefore, to expect a drop in the difference of life expectancy between men and women. It is doubtful that it will fall under 6.4 years by the end of the century. While the UN is being over-careful in its perspective concerning infant mortality and life expectancy for the population as a whole, it is perhaps being far too optimistic for the difference between the sexes. If the difference is effectively 6.4 years

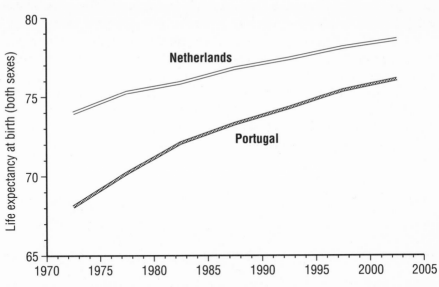

Figure 5.6 United Nation's forecasts of life expectancy at birth 1970–75 to 2000–2005 (EU countries with highest and lowest rates only).
Source: United Nations, *World population prospects* (1990).

in the year 2000, the average duration of life could be 75.5 years for men and 81.9 years for women. This difference would still be greater than that currently observed in Japan.

Finally we must consider spatial differences. Their evolution will probably follow the trend observed over the past decades, with a steady reduction in disparities. This reduction has been rapid, and may be expected to continue to be so during the 1990s (Figs 5.5 and 5.6). We might thus expect the maximum difference at the national level to be reduced by two years at the end of the century, taking into account the trend of homogenization within the EU. The average life expectancy in this case could reach 79.3 in the Netherlands and 77.3 in Portugal. These figures are slightly more favourable than those of the UN's Population Division. Variations at the regional level will remain greater but will also be diminishing.

It is, of course, important to note that these forecasts are being made for the European Union as it existed in 1994. With the enlargement of the Union by the accession of Austria, Finland and Sweden in 1995 the demographic situation of the new community of fifteen has changed from that of the twelve. In particular, the infant mortality rate in Finland and Sweden, at six deaths per thousand live births, equals the lowest national level within the EU, achieved in Denmark. Life expectancy at birth is also well above the EU average in Finland and Sweden. On the other hand, the accession of Austria brings in a country where the differential between

81

male and female longevity, at 7.2 years, is the second highest to France within the enlarged EU.

The overall prognosis must be that the European Union in the early twenty-first century (as with other advanced countries) will see the retention of strong mortality differences between social groups and between men and women, but that the spatial differences in mortality at both the national and regional scales will continue to be eroded. A greater degree of spatial homogeneity seems likely to emerge, although with certain differences being retained, for example class-based regional differences (as in parts of the UK), or cultural and dietary determined differences, as in parts of Belgium, France and Italy.

CHAPTER 6

The effects of the HIV epidemic on the population of Europe

MARKKU LÖYTÖNEN

Infectious diseases and population growth

The principle of population dynamics declares that it is environmental factors such as changes in physical conditions, food supplies and the occurrence of natural enemies that determine the number of an animal species in existence in a given area at a given point in time, and that these factors vary all the time, as the organic and inorganic environment is in a constant state of change. Every change that takes place in a species' living conditions is reflected in its numbers. Examined over a period of time, this takes the form of a constant search for a state of equilibrium, which in an undisturbed situation would result in a fluctuation around a certain average figure.

In the course of its endeavours to create a favourable environment for itself, human society has systematically attempted to prevent the spread of epidemics. Even so, the causes of such disasters were still unknown in the early nineteenth century, and it may be said that they were thus the last factors effectively limiting human population growth. Many of the advanced cultures of prehistoric times were obliged to carry on a bitter struggle for existence against the repeated threats posed by epidemics. Later the most serious outbreaks of the plague in Europe can be identified in the population growth curve for the continent in the form of distinct dips in total population figures. Similarly it is claimed that wars have often been more destructive through the accompanying spread of disease than in terms of the lives lost in the fighting itself (Cliff & Haggett 1988).

At the micro level, an imbalance comparable to the laws of population dynamics prevails between humankind and the micro-organisms that bring about diseases. Viruses, bacteria and other microbes are constantly altering their genetic make-up in search of new properties that will give them a hold over the human body, which has often become able to resist their attacks by virtue of its immune defence mechanism. This is a complex system that is able to acquire more effective means of combating disease agents which have attacked it previously. This means that a population which has had experience of a wide range of micro-organisms may be able to defend itself more readily than one which has been living in isolation. But it also means that a new cause of disease that appears from elsewhere will frequently lead to a serious epidemic on account of the population's lack of immunity to it.

When America was "discovered" in 1492, military power was scarcely needed to conquer the new continent, as far more native Americans died of the infectious diseases brought by the European soldiers than by the force of arms. And these native Americans in return gave the Europeans syphilis, which was a common but mild complaint in the Americas but which caused a serious epidemic lasting for several hundred years when it reached Europe. Like syphilis, cholera, influenza, tuberculosis and many other microbial diseases have served to control human populations for as long as the human race has existed (Ratledge et al. 1989).

The discovery in the second half of the nineteenth century of the micro-organisms responsible for these diseases was a scientific breakthrough that revolutionized our concepts of many human illnesses. It meant that the causes of infectious diseases could be pinpointed and progress could be made in their treatment, in the sense that for the first time in the history of humankind it was known what agent was responsible for the plague, for example. Since that time, improved hygiene, antibiotics, inoculations and many other basic techniques of modern medicine have proved effective in combating infections, perhaps the best example of which lies in the total elimination of smallpox as a result of the international vaccination campaign.

What are HIV and AIDS?

Although there are many infectious diseases that still present urgent problems in the Third World, the inhabitants of the industrial countries have for some time regarded themselves as protected from epidemics. After all, it is many decades since people at large were seriously afraid of catching cholera or the plague. Ten years ago, however, the dream of an endless triumph of modern medicine over dangerous infectious diseases was shattered. The appearance of patients suffering from AIDS (acquired immunodeficiency syndrome) in the early 1980s and the isolation a few years later of the HI virus (human immunodeficiency virus) that caused it reminded us that infectious diseases could still be a force to be reckoned with in the regulation of the populations of the industrial countries.

The HI viruses are lentiviruses, which were already known to cause slow, chronic infections in several animal species such as sheep. These in turn form part of the group of retroviruses, which are typically capable of leading to symptoms of disease a matter of years or even decades after infection.

The HI virus is passed from one person to another via contaminated blood products, contaminated medical instruments or shared injection needles, through sexual intercourse without protection, or perinatally. Once having entered the body, it attaches itself mainly to the helper T-cells, the key cells in the immune defence system. As it reproduces it gradually destroys these cells, and in this way weakens the individual's resistance to viruses, bacteria and other microbial infections.

The quantity of HI virus present in the body is large in the first few weeks after infection, as it reproduces rapidly while the immune system is unable to resist it, and some individuals can experience a primary syndrome lasting 1–2 weeks, or occasionally as long as a month. After this possible initial phase the subject is usually free or virtually free of symptoms and carries no external signs of HIV infection for a period which is thought at present to last some 5–10 years, although it is still possible to pass the infection on to others during this period. Gradually, in the course of the years, symptoms begin to appear, however, which indicate deterioration of the immune system, and microbes that would not seriously affect a healthy person can gain a hold. With the appearance of these opportunist infections the patient is regarded as having reached the final stage of HIV infection, the fatal condition known as AIDS.

WHERE DID THE HI VIRUS COME FROM?

As far as is known at present, the HI virus originates from East Africa where it, or some early prototype of it, first infected humans from three species of monkey found in the area. It was very probably endemic throughout the nineteenth century in the territory inhabited by these monkeys. No epidemic ensued, however, because African society was for a long time sufficiently static, with people's movements restricted to their own village or tribal community (Shannon et al. 1991, Ulack & Skinner 1991).

The situation altered after the Second World War, as the structural changes in society and the associated urbanization and industrialization led to greater mobility. It was at this stage that the HI virus began to spread to wider areas and the disease assumed the proportions of an epidemic in Africa (Barnett & Blaikie 1992). It is possible that individual travellers visiting that continent in the 1950s may have become infected, but these cases will not have attracted any particular attention from the health authorities, and any deaths that occurred must have simply remained aetiologically unresolved. Thus, the HIV pandemic developed unnoticed. The situation finally came to light in the early 1980s, when the number of cases escalated and the disease came to the attention of the health authorities in the USA.

How is it possible that the HI virus managed to spread beyond Africa to cause a pandemic extending over the entire inhabited world? The answer to this question is geographically a very interesting one. The biological regularities that maintain a state of equilibrium between a human being and the microbes that threaten him or her have developed to their present state over a period of 100–200 million years, and may be said to have safeguarded the existence and growth of human populations very well under the conditions in which the human species evolved. Populations were not large, and the density of settlement was not very great, in addition to which the world was for a long time inhabited by isolated human groups which had no contact with one another.

During the past hundred years, however, conditions for the occurrence and spread of epidemics have altered drastically. There has been what can only be described as a population explosion; people have moved to live in agglomerations of tens of millions residents; mobility has increased, and the speed of travel has become such that any point on the globe can be reached in a few hours. It is highly probable that the HIV pandemic would never have come to anything if human society itself had not altered its environment and ways of living.

The global HIV situation

How many HIV carriers are there in the world? This is a question for which there is no single answer that will satisfy everybody, for several reasons. The chief reason is that the World Health Organization (WHO) keeps statistics only of patients reaching the final AIDS stage. Some individual countries have statistics of symptomless HIV carriers as well, but these are not collected by the WHO.

Gaining a true impression of the HIV situation is made more complex by the extent of under-reporting, which is thought to amount to between 10 and 30 per cent in some of the industrialized countries, implying thousands or even tens of thousands of unrecognized sufferers in some places. A further problem is introduced by the fact that many of the developing countries are unable to compile any reliable data, or intentionally obscure the figures in order to give a more optimistic picture. When we know in addition that the symptom-free incubation period for the disease can last as long as ten years, it is obviously impossible to establish any figure for the number of people infected with HIV in the world.

In spite of these statistical difficulties involved, some estimates are available for the current number of HIV patients. The current WHO estimate in 1993 was about 13 million, but the situation is expected to deteriorate, and the same organization forecasts that there may be 30–40 million carriers by the year 2000.

Our knowledge of the geographical distribution of HIV carriers is also based only on isolated statistical data and estimates (Fig. 6.1). The most seriously affected continent is undoubtedly Africa, where the true number of carriers at the present time

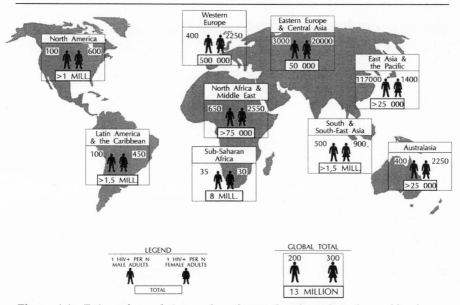

Figure 6.1   Estimated cumulative number of HIV infected people in the world and rate of infected by sex. *Source:* World Health Organization (1993) and United Nations, *Demographic yearbook* (1991).

is likely to run into millions. It should be remembered, however, that regional differences within the continent are considerable and that it is impossible to obtain reliable data from all countries. The situation is most serious in East Africa, around Lake Victoria, where as many as 30–40 per cent of the population are infected in some places. On the other hand, the state of the disease in the Islamic countries of North Africa is almost entirely a matter of guesswork, as it is extremely difficult to obtain any reliable information (Smallman-Raynor et al. 1992, Gould 1993).

The next highest prevalence is recorded for North America and South America, where there are probably over a million HIV carriers on each continent. There are five states in the USA, for instance, where AIDS is the leading cause of death among men aged 22–44 years. In the third rank come Europe and South and Southeast Asia, but here again there are substantial regional variations. Other parts of the world are found at the bottom of the table of occurrences, but this is perhaps partly because the true situation in these areas is not reliably known.

ONE PANDEMIC – MANY EPIDEMICS

From a geographical point of view, the HIV epidemics in the various continents appear to be developing in different ways (Mann 1991). It is characteristic of the situation in North and South America and in Europe that most cases concern

87

homosexuals and those taking drugs intravenously. It is also evident in the USA that there are proportionally fewer carriers among whites than among other sectors of the population. In Africa, on the other hand, infection is just as common in women as in men and spreads mostly through heterosexual relations. Attempts have been made to explain this by the fact that the common venereal diseases also serve to promote the spread of the HI virus, but it is also obvious that the general development problems prevailing on that continent, the shortage of economic resources, and the poor standards of health care go a long way towards accounting for the epidemiological differences between the African countries and those of the western industrialized world.

The situation in Asia appears to be converging rapidly with that in Africa. At first it seemed that the HIV epidemic had begun more slowly in Asia, but this was evidently an illusion, and the true situation has proved to be at least as serious. In fact the numbers of carriers have increased extremely rapidly, and it is feared that the epidemic may gain momentum through such effects as prostitution.

Existing epidemiological models indicate that the HI virus will eventually spread more generally to the heterosexual population in the western countries as well, a conclusion based on the fact that this is the predominant pattern in Africa, where the epidemic is most advanced. There are now, in fact, some signs of this happening, which will prove the predictive models to be correct. Also, there are no medical reasons why the epidemic should restrict itself only to a certain subpopulation, such as homosexuals or intravenous drug abusers.

On the other hand, it is quite possible that the epidemics in industrialized countries and in less developed countries may continue to differ in character. Prevailing differences in the epidemics can be seen as being a result of various social and economic factors. Poor health care systems, lack of resources for health campaigns, low levels of education – all well known indicators of underdevelopment – widespread prostitution, the high prevalence of sexually transmitted diseases, and political instability have all contributed to the rapid diffusion of the HI virus in many parts of Africa and Asia. It is possible that more than half of all HIV infections in the world are more or less connected to poor social and economic conditions.

The vast number of clinical and laboratory orientated research projects dealing with HIV and AIDS have already pointed out that we are facing a medically challenging problem. It seems likely that it will take another decade until the final secrets of the fatal virus have been revealed and we have any means other than information campaigns for intervention in the epidemic. Even if a curative treatment or effective prevention (for example, through vaccination) became available in the near future, the resources needed to stop the epidemic may well exceed the national health budgets available in most of the countries of the less developed world. In the industrialized world, however, it is more likely that prevention and treatment, once developed, will be made available for use regardless of the costs.

AIDS and HIV in Europe

It is easier to form an accurate picture of the situation regarding HIV infection in Europe than it is in Africa or Asia. As elsewhere, the main European statistics concern only AIDS patients, these being compiled by the World Health Organisation and the joint World Health Organisation–European Community Collaborating Centre on AIDS, published quarterly in the form of a comprehensive statistical review available for use in all countries. The statistics are naturally affected by the same problems as elsewhere in the world: underestimation, deficient data collection, temporal discontinuities caused by political changes, and the general unreliability of the data for eastern Europe in particular. Although the WHO estimate is that there are about half a million HIV carriers in Europe, any factual consideration of the epidemiological situation still has to be based only on reported cases of AIDS, since only a few European countries maintain systematic statistics on the prevalence of HIV infection. Thus, it is important to remember when looking at the maps (Figs 6.2, 6.3) that the data used refer only to actual cases of AIDS and that they therefore give only an approximate picture of the real state of the HIV epidemic in Europe.

A total of 92,769 cases of AIDS had been reported in Europe by the end of March 1993 (Fig. 6.2), but the situation in regional terms was highly variable. Cases had been reported in all countries except for Albania, Azerbaijan and Kazakhstan, with the highest absolute figures recorded in France, Italy and Spain. In fact these three countries together accounted for over 60 per cent of all AIDS cases in Europe, the remainder occurring predominantly in the other countries of western Europe. The only eastern European country with comparable figures is Romania, where the majority are paediatric patients who have received the infection through deficient hygiene and staff negligence during hospital treatment.

The situation changes very little when the prevalence of AIDS cases is examined relative to the total population (Fig. 6.3). Ignoring Monaco, with its minuscule population base, Spain now emerges at the top of the infection list, followed by Switzerland and then France and Italy in that order. The distribution is still weighted towards western Europe.

There are grounds for suspecting that the reporting of AIDS cases may be less reliable in eastern Europe, but although it may seem necessary to take account of the political instability of some of these countries, the general impression provided above may well be largely correct. It is quite possible that the decades of isolation experienced by the eastern European countries may have effectively protected them from the spread of the HI virus. Although the numbers of visitors to these countries amount to some millions every year, the restrictions placed on foreign travel by their own citizens in the past will have kept the number of cases of HIV infection low, and the restrictions on internal movement, especially in the area of the former Soviet Union, will have prevented any spread of the virus within the

89

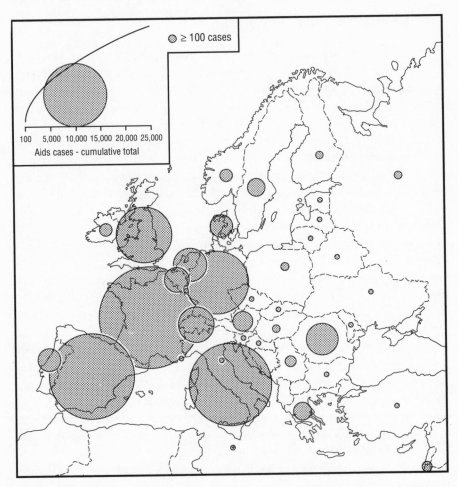

Figure 6.2   Cumulative number of reported AIDS cases in European countries by 31 March 1993.
Source: AIDS Surveillance (1993).

country. Support for this argument is obtained from the fact that the paediatric cases that have arisen in Romanian and ex-Soviet hospitals on account of poor standards of hygiene and low working morale have remained very local in scale.

The curves for the increase in AIDS cases in different countries are very similar in appearance. The number of cases increased in more or less an exponential manner in the first few years, and the first mathematical forecasts were based on these figures, thus predicting a rapid increase in all parts of Europe. The trend had begun to even out slightly by the late 1980s, however, suggesting that this is not such a rapidly advancing epidemic as was first feared. The latest prognoses are that the number of AIDS cases will continue to increase all over Europe but only at a mod-

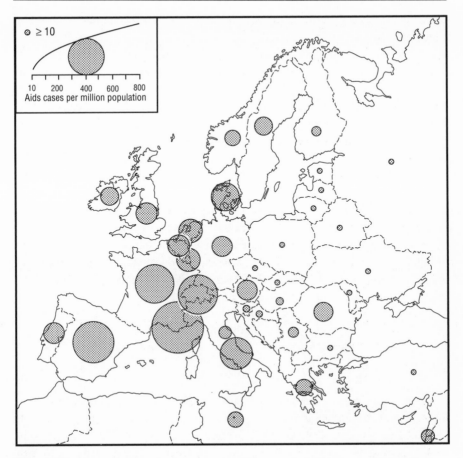

Figure 6.3 Rate of reported AIDS cases per million people in European countries by March 31 1993.
Source: AIDS Surveillance (1993).

erate rate. There will be no explosive spread as had been originally forecast. The AIDS predictions for England and Wales, for example, have just been adjusted downwards to a considerable degree relative to those conjectured some time ago.

There are several reasons for these more optimistic forecasts. In the first place, new cases of AIDS were being discovered rapidly in the early and mid-1980s, representing HIV infections that must have taken place between the late 1970s and the early 1980s. Secondly, HIV tests were available on the market by the mid-1980s, and these facilitated the reliable determination of whether or not an individual was an HIV carrier. Thirdly, the systematic testing of blood donors and the antibiotic treatment of blood products virtually eliminated infection from that source in Europe. And fourthly, all the European countries are committed to a campaign against behaviour likely to promote the risk of HIV infection. All this is reflected in the figures for individual countries, showing a sharp early increase in cases, followed by a reduction in the rate of increase.

The trend in eastern European is to a great extent similar to that in the West, and although the data are both incomplete and to some extent unreliable, an increase in the number of cases is certainly to be seen. For the reasons mentioned above, however, the eastern European countries are lagging a few years behind, so that where most countries in the West are currently talking of hundreds or thousands of AIDS patients, those in the East often amount to less than a hundred in each country.

What, then, is the true situation in Europe? Does the picture given above provide a fair impression of the geographical distribution of the estimated half a million carriers and of the state of the HIV epidemic? It is very difficult to answer this question directly without simply giving a list of pitfalls and reservations. It is probable, however, that in spite of the statistical deficiencies, the systematically compiled data on AIDS cases inform us fairly reliably of the state of the HIV epidemic in Europe at the present time. Its spread continues to be more advanced in western Europe, with the countries of eastern Europe lagging a few years behind.

CAN THE EFFECTS OF THE HIV EPIDEMIC BE RELIABLY PREDICTED?

Epidemiological forecasting and spatial analysis of the diffusion of contagious diseases are essential methods for the monitoring and controlling of epidemics. It involves both geographical and mathematical modelling. In addition to different geographic and demographic variables, the data needed in such modelling deal with disease-specific factors providing fundamental information for estimating the necessary parameters. In most diseases these parameters are well known, and can be reliably estimated from existing data or can be obtained by analyzing a representative sample of case histories. Based on such data, modelling of the growth curves and the spatial diffusion processes can be done with a high degree of accuracy. In such a case, it is also possible to forecast the impact of a disease on the population in a given region.

Ever since AIDS patients were first so identified in the early 1980s, the forecasting of the HIV pandemic on international and national levels has been one of the main goals of all epidemiological research. When attempting to model the HIV epidemic, however, the situation is more complicated when compared with, for example, an influenza virus pandemic. While clinical, virological and immunological research has progressed rapidly, epidemiological research on HIV has been riddled with the data-related problems mentioned earlier. Estimates of the growth of the HIV epidemic have proved to be far less successful than one might wish (Löytönen 1991).

The problems in forecasting the HIV epidemic are mostly attributable to the virological features of the virus, and its connection with various forms of human sexual behaviour, in other words to some of the more hidden sides of human life. Although many of the fundamental disease-specific medical factors controlling transmission at the individual level are still unknown, there are several documented cases in which transmission has resulted from a single exposure. On the other hand, there are several equally well documented cases where no transmission has occurred despite repeated exposure with the same or different carriers.

In terms of forecasting the epidemic from a geographical standpoint, most of the factors controlling the course of the epidemic in a given population or region are poorly known. These factors include, among others, the mean and group-specific incubation time; the rate of risk associated with different means of transmission; the number of partners and frequency of sexual intercourse and the variation in these indicators in the heterosexual, bisexual and homosexual populations; the risk of being infected *in utero*, and the role of transmission co-factors such as genital ulcers.

When AIDS was identified in the early 1980s it was first documented in homosexuals. Although HIV infection is not limited to this group, their life-style has in the past often been prone to the acquisition of sexually transmitted diseases. In relation to the transmission of the HI virus among homosexuals, it is well documented that the number of sexual partners is positively correlated with the risk of acquiring the HIV infection. This is partly attributable to the fact that HIV transmission is more likely to occur with anal than with vaginal intercourse.

Although first found among homosexuals in the industrialized countries, both bisexuals and needle-sharing intravenous drug abusers most probably spread the virus into the heterosexual population. Various of these subgroups overlap, often linked through prostitution, whereas in some countries individuals from these groups may have been active blood donors for economic reasons. Contaminated blood products have played some role in spreading the HI virus in the early stages of the epidemic in the industrialized countries. Today, processed blood products are reliably tested for the HI virus prior to use in the western world. Recent news stories from France and Germany, however, show that contaminated blood products can still play a minor role in spreading the virus. In epidemiological terms, surgical interventions are insignificant in spreading the virus, although poor standards and lack of quality control in the health-care systems of the less developed world often leave much to be desired. Even if no curative treatment becomes available, it is possible that the growth curves in the industrialized countries will level off as a result of massive information campaigns supported by well organized and efficient health-care systems, which can identify infection at an early stage. Such progress seems less likely to occur in the less developed parts of the world.

As already pointed out, the data problem is such that most attempts at forecasting the HIV epidemic are based on data covering only symptomatic AIDS patients. From

93

the epidemiological point of view, the primary focus should always be on analyzing the spread of a virus, and not merely on the statistics concerning the clinical development of the infection – which again varies much depending on several factors such as the availability of clinical treatment and the quality of the health care system in general.

Until complex statistical and mathematical epidemiological models and their geographical extensions and modifications become more successful in forecasting the future course of the HIV epidemic, evaluating the impact of HIV on Europe's population as a whole should be done on a more qualitative basis. Although it is possible to produce fairly reliable forecasts on a national level for some European countries, such as Finland (Löytönen 1991), an attempt at such statistical forecasting for the whole of Europe would almost certainly lead to an unreliable result.

The future of the HIV epidemic in Europe

The probable trend in Europe is that the HI virus will spread geographically to all parts of the continent but that the epidemic will alter in character socially to become an infection transmitted through heterosexual behaviour in the same manner as the other venereal diseases. The differences between western and eastern Europe are likely to even out gradually, possibly before the year 2000, largely on account of the fact that the increase in the numbers of both HIV carriers and AIDS cases would seem to be slowing down in the West at the same time as the epidemic is just getting under way in the East. One is perhaps entitled to be sufficiently optimistic, however, as to expect both figures to settle eventually at some sort of average level around which the annual statistics will fluctuate, this situation probably being reached somewhat later in the East. If this "peaceful development" scenario comes about, we can estimate that the effect of the HIV epidemic on demographic trends in Europe as a whole will be no more than a marginal one over the next 10–20 years, although it may vary greatly from one country to another.

This scenario can be justified on several grounds. Perhaps the most telling of these is experience regarding the spread of other sexually transmitted diseases, the incidence figures for almost all of which have settled at a certain average level characteristic of the disease concerned, with some annual fluctuation around this average. It is obvious, of course, that some changes in the incidence of diseases take place in the course of time (for example, syphilis and gonorrhoea are declining in most countries whereas trachoma is on the increase), but such changes usually take place slowly, over a matter of decades.

A second justification is that as a result of the large-scale, intensive programme of research, we now know a great deal (although by no means all) about the HI virus, its manner of infection and its progression in the patient. On the other hand, it should be remembered that, in spite of many attempts, no specific form of treatment has been devised which is capable of curing a person carrying the HI virus, and even the work of developing a vaccine to halt the epidemic has made very slow progress. It is probable that substantial resources will continue to be devoted to research into AIDS and the HI virus in the future, and this gives cause for a certain optimism. It is conceivable that more powerful drugs will be found that will help HIV carriers, and it is also possible that the vaccine tests will eventually lead to the discovery of a sufficiently effective means of stopping the spread of the epidemic. It will be many years before this is achieved, however, perhaps running into decades.

The third justification concerns patterns of human behaviour. If the numbers of HIV carriers continue to increase at an accelerating rate, as was forecast at the beginning of the epidemic, people will react to the situation and change their habits. The education campaigns organized throughout the industrialized countries and the prominence give to the subject in news broadcasts and media discussion have already ensured that most people in Europe have heard of AIDS and the HI virus. Should the epidemic spread at an accelerating rate in Europe and other parts of the world in spite of all the efforts made to contain it, it is highly likely that the fear of infection will drive people to alter their sexual behaviour and adopt safer practices.

What are the average figures at which the incidences of HIV infection and AIDS cases in the European countries will settle, and how will the virus spread geographically in the future? On account of the problems entailed in the compiling of statistics and mathematical models, as discussed above, it would be unreasonable to try to quote precise figures to show how the spread of the virus is likely to affect population growth in Europe or its individual countries. Joint research carried out in certain countries by epidemiologists and geographers nevertheless does permit relatively reliable short-term predictions to be made. The statistics on AIDS cases in western Europe provide at least an adequate basis for this, and some good examples of such predictions are already available, such as in Finland (Löytönen 1991).

THE EPIDEMIOLOGICAL DICHOTOMY IN EUROPE

The "peaceful development" scenario quoted above could obviously be disturbed at any time, for several reasons Easterbrook & FitzSimons 1992, Löytönen 1992). Perhaps the greatest uncertainty factor and the most difficult to evaluate concerns developments in eastern Europe in the near future. The region is at present in a politically, economically and socially unstable condition. The collapse of the Soviet system and the subsequent rapid liberalization revealed these former socialist countries as being in a state of social and economic depression and backwardness. Damage to the environment has proved to be far more serious than expected, and far more difficult to rectify, and the training of a skilled labour force and initiation of the research and development work necessary to maintain competitive industries is retarded by a shortage of capital.

The heavy industry built up under the socialist system as the backbone of the economy in this region is hopelessly outdated and only serves to compound the environmental problems. In practice, it will be necessary to construct, more or less from nothing, new industries based on well disciplined logistics and flexible production systems. It will take a long time to eradicate the air of indifference that permeated the social atmosphere and industrial morale of the Soviet period, and this will call for a great deal of re-education and perhaps the emergence of a whole new generation. In spite of the economic assistance provided by the West, it will be several decades, even with the most optimistic outlook, before the countries of eastern Europe catch up with their western neighbours in terms of living standards.

At the political level, the former socialist states are now trying to distance themselves from the structures of the communist era both in practical terms and psychologically, by emphasizing their independence and sovereignty. It is a process that seems to be directly opposed to the increasingly powerful tendencies towards integration among the countries of western Europe. This suggests that the abrupt step from a national philosophy built on sovereignty to a mood of integration motivated by a desire for economic advance will probably not be achieved particularly easily or quickly in the former socialist countries – especially since it is evidently not proving a straightforward matter even for the western countries themselves.

As far as the HIV epidemic is concerned, there are risks connected with the set of problems just described. Economic, social and political unrest provides an ideal breeding ground for increased crime, drug abuse, prostitution and in general the kind of human behaviour that points to an unhealthy society. Those who engage in these activities do not necessarily regard them as short-cuts to affluence, of course, but as a means of gaining some sort of living under the prevailing conditions, possibly the only means open to them. If economic growth in these countries cannot be quickly stimulated – and there is little hope of that as the Western industrialized countries are struggling with their own recession – it is likely that the present relatively favourable situation with regard to HIV in eastern Europe will deteriorate rapidly. Since the optimistic scenario described earlier suggests that the situation in western Europe will soon have advanced to a stage where the rate of increase in the numbers of carriers slows down, the worsening of the situation in the East may well place new pressures on the West and lead to a new wave of the epidemic there. Migrant workers coming to the West in search of higher incomes could well bring about new infections, as also could prostitutes from the eastern European countries, who are already to be found all over western Europe. The division of Europe into "epidemiological camps" of East and West is thus one notable feature that could well threaten the assumptions of the "peaceful development" scenario.

RICH VERSUS POOR COUNTRIES IN THE HIV EPIDEMIC

The treatment of HIV carriers is very expensive, especially since monitoring and therapy is required throughout the germination period, possibly of more than ten years, while the further the infection progresses, the greater is the amount of treatment required and the more expensive it becomes. It has been estimated in the USA, for example, that AIDS patients will be consuming about a fifth of the money spent on health services by the end of this decade.

The richer industrialized countries may well be able to cope with the costs of the HIV epidemic, but the situation in the developing countries is far more serious. The treatment of HIV carriers is already eating into the limited resources available for health services, while the fact that most of the people dying of AIDS in Africa, for instance, are in their most productive working ages means substantial financial losses for their families and for society at large. The gap is enormous between the poor and rich countries in terms of their chances of overcoming the economic problems raised by the HIV epidemic, regardless of whether the financial sums concerned are related to the total population or to the number of HIV cases.

The worldwide vaccination campaign to eliminate smallpox was successful because the basic costs (vaccine, instruments and personnel) were low, the project gained broad international acceptance and the vaccination procedure did not require any advanced professional skills. Such a campaign could in principle be mounted against the HI virus once a vaccine has been developed that gives efficient protection. Current information nevertheless suggests that any vaccine that might become available, whether it provides full immunity or merely retards the infection in the body, is likely to be a very expensive preparation even by the standards of the industrialized countries and to require advanced medical techniques and a highly trained health-care staff for its administration. This means that a vaccination programme would be so expensive to implement that the meagre health service resources of the developing countries could not bear the strain, and finance could never be obtained for a worldwide effort on the scale of the smallpox campaign. The outcome could very well be a situation in which the industrialized countries are able to control the advance of the HIV epidemic, and in the best case arrest it completely, by the techniques of modern medicine combined with the expenditure of the necessary funds, whereas the ever more rapidly growing population of the developing countries will have to face an ever more serious HIV situation.

Could the worsening epidemic in the developing countries pose a serious threat to the population of the industrial countries, for example in Europe? Even a cautious evaluation of the situation suggests that it is quite possible that the growing population potential of the developing countries will eventually result in increasing migratory flows exerting pressure on the borders of the industrial countries – indeed, there are already signs of the attractiveness of these countries in the eyes of

Third World populations. Such migrants will naturally bring with them the diseases that are common in their own countries. Human history already shows experiences of how infectious diseases can cross the world with people who are on the move.

Is the HIV epidemic a sign of the limits of population growth?

Fundamentally, the world's human population is dependent on environmental factors in the same way as are the populations of other animal species. Human biological evolution has nevertheless given us one property which raises us above all other animals: we are the only species capable of inheriting knowledge cumulatively produced by previous generations. All other species can only inherit in genetic form. The resulting unprecedented human capacity for amassing and using information has created circumstances in which we are able to manipulate our environment and to render it more favourable for the growth of the species. The result has been a rapid population explosion, which has now reached the point where it is causing serious disturbances in the balance between humankind and nature. No permanent state of equilibrium can be perceived for the immediate future, and current signs are that the world population will simply go on growing at an increasing rate.

Although the population of Europe now seems to be safe from the threat of the HIV epidemic, it is indeed justifiable to ask the rhetorical question of whether the HI virus – which is, after all, relatively slow to migrate – is a first sign that human population growth is reaching its limits. Unrestricted growth on the part of one species is simply not possible in a restricted world.

CHAPTER 7

Internal migration, counterurbanization and changing population distribution

TONY CHAMPION

Introduction

The geography of population change in Europe has altered markedly in recent decades, with some remarkable switches in the location of the areas of fastest growth
between the 1950s and the 1970s and with a rather general decline in the scale of
geographical variations in growth rates since then. The dramatic declines in fertility
and rates of natural increase in southern Europe (see Ch. 2) have played a major role
in these trends, as has the long-established tendency towards convergence in mortality rates across Europe (see Ch. 5). Nevertheless, at the regional level it is the
migration component that has been primarily responsible for the developments in
population distribution.

Both international and internal migration have altered significantly in their
nature and patterns since the Second World War. As regards international
exchanges, Europe has switched from net exporter to net importer of people, and
the predominant type of migrant process has shifted over time from the guest-
worker phenomenon of the 1960s and early 1970s to family reunification and most
recently to what White (1993b) terms "post-industrial movement" comprising
skilled labour, illegal migrants and asylum-seekers. In the 1950s and 1960s internal
migration was dominated by rural-to-urban migration, along with increasing concentration of population in the capital regions of most countries, but the 1970s has
become known as the decade of counterurbanization as centrifugal tendencies
grew in strength.

Anticipations of future developments are clouded by the volatility of migration
in general and the rather indeterminate character of internal migration patterns in
the 1980s. In particular, the late 1980s and early 1990s have seen the large-scale
influx from eastern Europe and the continuing build-up of population pressures
from the South. Meanwhile, the level of net internal movement was considerably
smaller in the 1980s than in the three previous decades and, as a general rule, did not
follow clearly either the counterurbanization pattern of the 1970s or the strong concentration pattern of the 1950s and 1960s.

99

While the next three chapters focus on different aspects of international migration, this one concentrates on the emerging patterns of internal movement. It begins with an overview of the population redistribution trends that have affected Europe since mid-century, looking in particular detail at the counterurbanization experience and the extent to which this has backtracked since the 1970s. It then reviews the various explanations that have been put forward to account for these past trends in order to provide a framework for anticipating how patterns may develop over the next few years. It finishes by attempting to look into the future by discussing how two major elements of past population redistribution – economic restructuring and population deconcentration – may evolve over time and then by examining two examples of scenario-based exploration.

The 1980s in context

The primary purpose of this section is to profile the extent and nature of spatial variations in population growth over the past decade and to set these in their longer-term context in order to discover any trends which seem consistent enough to continue into the future. A brief look is taken at the national-level picture, highlighting the narrowing of inter-country differences in growth rates. Then, focusing on the regional map, the role of the core–periphery and urban–rural dimensions is explored. This is followed by a statistical assessment of the relative importance of concentration and deconcentration tendencies.

Because of the considerable problems of obtaining regional-level data on past population trends for eastern Europe, this account deals only with Europe west of the former Iron Curtain. This comprises the member states of the Council of Europe as of 1991, thus including Turkey but excluding the former German Democratic Republic as well as former Czechoslovakia, former Yugoslavia, Poland, Hungary, Roumania and Bulgaria. Apart from the inclusion of Turkey, the coverage is very similar to the European Economic Area, as made up by the European Union and European Free Trade Association. The data are drawn largely from Council of Europe publications, together with information supplied by the statistical offices of member countries for a report commissioned by the Council of Europe (Champion 1991).

THE NATIONAL-LEVEL CONTEXT

As much of the change over time in the regional map of population change can be linked to national and international demographic trends, it is useful to provide at least a brief overview of these. Four aspects are particularly relevant: the general reduction in overall growth rates, the narrowing of inter-country differences in growth rates, the waning of the contrast between Mediterranean countries and the rest of Europe, and the reduction in the number of net emigration countries.

The general reduction in overall growth rates has been noted in previous chapters and is well illustrated by historical data on the European Union area as it stood on the eve of the 1995 enlargement. The twelve countries, in aggregate, saw their annual average growth rate fall progressively from 0.96 per cent in 1960–65 to 0.71 in 1965–70, 0.61 in 1970–75, 0.37 in 1975–80 and 0.27 in 1980–85. The latest five-year period 1985–90, however, shows a departure from this trend, with the rate moving up to 0.34 per cent (Eurostat 1991a).

The reasons for these developments can be identified most directly by reference to data on the change components. They have nothing to do with the crude death rate, which has changed hardly at all over the past 30 years, as the effect of the continued fall in mortality rates has been countered by population ageing. The general reduction in overall growth rates is very largely attributable to the change in crude birth rate, which has fallen through each successive five-year period from 18.8 per thousand in 1960–65 to 11.8 in 1985–90. The recent upturn in growth rate is entirely attributable to a rise in net immigration, up from 0.5 per thousand in 1980–85 to 1.6 in 1985–90 (Eurostat 1991a).

Secondly, the narrowing of inter-country differences in growth rates is a long-established trend but has progressed particularly strongly since the 1970s. Concentrating on Council of Europe member countries with at least one million people but excluding Turkey, the average annual rate of growth in the 1950s ranged from 1.42 per cent to −0.46 per cent, a difference of 1.88 percentage points. The range narrowed to 1.68 points in the 1960s and to 1.43 points in the 1970s, but then shrank even more rapidly through to the 1980s when it stood at 0.83 percentage points according to data provided by Hoffmann-Novotny & Fux (1991).

The third theme is the waning of the contrast in demographic profiles between southern Europe and the rest of the continent. The population of Portugal, Spain, Italy and Greece grew, in aggregate, by only 3.5 per cent between 1980 and 1990. This was not very much higher than the level of 2.9 per cent observed for the central and northern member states of the Council of Europe in the 1980s. The differential had been much greater in the previous decade, when the rates had been 7.4 per cent and 3.3 per cent respectively. Then the significantly higher figure for southern Europe reflected its higher birth rate, as well as some degree of return migration from northwest Europe following the 1973/74 economic recession and breakdown of the guest-worker system.

Lastly and partly related to this is the marked reduction over the past 20 years in the number of countries recording net emigration. In 1970 nine of the larger Council of Europe member countries lost population through international move-

101

ments: Finland, Greece, Ireland, Italy, Norway, Portugal, Spain, Switzerland and Turkey. In 1980 the number had shrunk to five, and by 1991 it was down to just one: Portugal. The upsurge in migration pressures from eastern Europe and the Third World was making its mark by the end of the 1980s, reinforcing the cutbacks in labour recruitment by northwest Europe and the reduced level of the European exodus to the New World.

REGIONAL PATTERNS OF POPULATION CHANGE IN THE 1980S

Some basic indicators of regional population change for the largest 16 Council of Europe member countries are presented in Table 7.1. The most impressive feature is the wide range of regional growth rates, despite the fact that the degree of inter-country variation was much lower in the 1980s than it had been previously. This is true not only across Europe as a whole but also within the individual countries. The largest range is found in Portugal, with both the fastest growing region and the fastest declining one in Europe (apart from the case of Turkey), but even the country with the narrowest regional range (Belgium) exhibits a wider range than exists between the national growth rates described above.

Table 7.1 also shows the regional range of crude birth rates and migratory change rates for each country. The former reflect the impressive changes which have occurred in southern Europe in recent years, with some of the regions in Italy, Spain and Portugal being characterized by the lowest birth rates of all the 385 regions included in the analysis. Migration, however, is the dominant component of population redistribution between regions. For virtually all the countries for which comparisons can be made, the differences between regions in annual average net migration rates are considerably larger than those for the birth rate. In the case of Belgium, for instance, the range of 9.8 points between the highest and lowest regions for net migration rate is more than three times that for the birth rate, and in several other countries the former is around double the latter.

The regional map of net migration change rates in the 1980s (Fig. 7.1) reinforces these points. In particular, it shows that the degree of variation within countries totally eclipses any broader geographical variation across the continent. Certainly there would appear to be no clear gravitation of migrants towards the European "core" of northwest Europe. Areas of migration gain and loss are scattered widely across the map.

Moreover, the overall impression conveyed by Figure 7.1 is markedly different from the dominant periphery-to-core population shifts that took place within countries in the 1950s and 1960s. While the latter would still appear to be a fair description of regional patterns in Scandinavian countries, the geographical

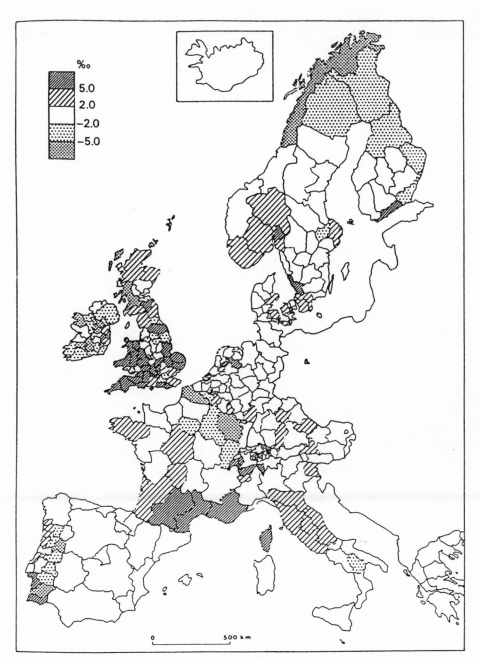

Figure 7.1 Average annual rate of net migration per thousand people in the 1980s. (See Table 7.1 for details of the specific periods used for each country. Note that there are no data for Greece or Spain.
Source: after Champion (1991).

Table 7.1 Range of population change indicators between "regions" within countries in the 1980s.

Country	Statistical areas	Population change			Birth rate			Net migration rate		
		Min	Max	Diff	Min	Max	Diff	Min	Max	Diff
Austria	Bundesland (9)	−4.5	+7.3	11.8	10.1	13.7	3.6	−1.6	+2.7	4.3
Belgium	Province (11)	−3.3	+6.3	9.6	10.9	14.0	3.1	−3.8	+6.0	9.8
Denmark	Amt (14)	−8.3	+8.0	16.3	−	−	−	−1.7	+5.3	7.0
Finland	Province (12)	−3.1	+10.8	14.1	10.8	14.6	3.8	−3.4	+5.9	9.3
France	Region (22)	−1.0	+13.3	14.3	9.7	15.8	5.3	−6.3	+12.6	18.9
Germany (west)	RB (31)	−7.2	+8.6	15.8	9.0	12.0	3.0	−1.9	+15.2	17.1
Ireland	County (26)	−7.1	+23.1	30.2	−	−	−	−7.0	+5.0	12.0
Italy	Region (20)	−5.1	+7.2	12.3	6.5	14.2	7.7	−2.8	+4.6	7.4
Netherlands	Provincie (11)	+0.7	+13.8	13.1	11.3	14.0	2.7	−2.2	+6.8	9.0
Portugal	Distrito (20)	−8.6	+26.9	35.5	9.7	15.4	5.7	−7.0	+21.9	28.9
Spain	Communidades auto (16)	+0.8	+11.7	12.5	8.3	14.3	6.0	−	−	−
Sweden	County (24)	−3.7	+8.9	12.6	11.4	14.5	3.1	−4.1	+7.0	11.1
Switzerland	Canton (25)	−1.0	+14.6	15.6	8.7	17.5	8.8	−4.6	+8.5	13.1
Turkey	Province (67)	−7.7	+48.7	56.4	−	−	−	−23.0	+12.3	35.3
UK	County (69)	−7.0	+14.5	21.5	10.5	20.4	9.9	−8.0	+18.9	26.9

Note: All data are in annual rates per thousand.

Periods: Population change relates to 1980–88, except for France, Germany and Spain 1980–87. Ireland 1981–6; Turkey 1980–85; United Kingdom 1981–8. Birth rate is for 1988, except for Belgium, France, Germany, Spain, United Kingdom 1987. Net migration rate for 1980–88, except for Belgium, France, Germany, Italy, Spain 1980–87; Ireland 1981–6; Turkey 1980–85; United Kingdom 1981–8.

Areas: In Belgium, Brabant is subdivided into Brussels, Dutch-speaking, French-speaking. In Netherlands, Flevoland is grouped with Overijssel. In Switzerland, Basel-Stadt is grouped with Basel-Landschaft. In United Kingdom, local authority regions are used for Scotland, Boards for Northern Ireland.

Source: calculated from data supplied by Eurostat and the national statistical agencies. Reprinted from Champion (1991).

patterns have generally become more difficult to interpret, appearing to owe more to special factors operating on a localized or short-term basis than to any single all-important process.

Nevertheless, some broad generalizations can be made. Regions experiencing net out-migration include two rather distinct groups. One comprises heavily urbanized areas, mainly in northwest Europe and including most of the capital cities (London, Paris, Brussels, Copenhagen), with the most severe losses occurring where urban decentralization is reinforced by industrial decline and the decline of port-related activities. The second is composed of less developed low-income rural areas, particularly in the rural hinterlands of the larger agglomerations in southern Europe but including the less heavily populated areas of Ireland and Scandinavia.

Net in-migration regions form a more disparate set, but at least three types can be identified. One comprises the hinterlands of the major cities, mainly in northern and central Europe, that are benefiting from the urban exodus, such as from London, Copenhagen and Randstad Holland. Gains are also being made by many of the "sunbelt" zones, these usually being less urbanized areas with medium-sized and smaller cities, most notably southern areas of the UK, Germany, France and Portugal together with central and northeastern Italy. A third category are some of the larger urban centres themselves and/or their immediate hinterlands, particularly in Scandinavia, Mediterranean Europe and Ireland.

In summary, with the narrowing of birth-rate differentials across Europe, the regional map of population change has become increasingly dominated by migration exchanges. Despite the recent switch of many countries from net losers to net gainers of international migrants, however, levels of net interregional movement in Europe have been lower in recent years than they were in the 1960s and early 1970s, and the patterns are less clear-cut than in the past. While the existence of major urban centres exerts a powerful influence on migration patterns, their role appears to vary between – and, in some cases, within – countries.

TRENDS IN REGIONAL POPULATION CONCENTRATION AND DECONCENTRATION

The particular aspect of migration which is raising so much uncertainty at present is the recent experience of "counterurbanization" and, more generally, trends towards regional concentration and deconcentration. As mentioned above, the first half of the postwar period was dominated by the movement of people from peripheral regions to national core regions, associated with widespread rural–urban migration and universal increases in the proportion of the population living in urban areas. The 1970s, however, seem to have represented a significant departure from this pattern, as many of the larger cities and more densely populated regions experienced migration losses, but since then most countries have not seen any acceleration of this process but instead have registered a slowdown in such deconcentration or indeed a return to larger-city migration gains.

The strength of urbanization tendencies in the 1950s and 1960s is clearly attested by official statistics. According to the United Nations (1992) the proportion of people residing in urban places in Europe as a whole grew from 56.2 per cent in 1950 to 66.6 per cent in 1970. For the European Union, the respective figures were 64.8 and 74.0 per cent, and all twelve countries participated in this trend, albeit to differing extents. Those with already high levels of urbanization in 1950, such as

105

Belgium (91.5%), the UK (84.2%) and the Netherlands (82.7%) saw increases of only 3–4 percentage points over the 20-year period, in contrast to France, Greece and Spain where the levels rose by around 15 percentage points (Champion 1993).

Since 1970 two changes have occurred. In the first place, while the proportion living in urban places has continued to rise, it has done so at a considerably slower rate. By 1990 the level of urbanization had reached 73.4 per cent for the whole of Europe and 78.9 per cent for the European Union area. Secondly, over the past two decades, people have been redistributing themselves down the settlement hierarchy. According to the World Bank (1992) the proportion of the urban population accounted for by cities of one million people or more fell in several countries between 1965 and 1990: for instance, down from 33 to 26 per cent in the UK, from 38 to 31 per cent in Denmark, from 42 to 37 per cent in Italy, and down by 4 percentage points in Austria, France and Greece.

This waning of urbanization tendencies has already been well documented for the 1970s (see, for instance, Hall & Hay 1980, van den Berg et al. 1982, Fielding 1982, Vining & Pallone 1982, Cochrane & Vining 1988, Illeris 1988, Champion 1989). The larger cities, notably those which began their rapid growth in the nineteenth century, were generally observed to be much less attractive for migrants in the 1970s than in the previous two decades and, in many cases, were experiencing substantial migratory losses at this time. Moreover, the scale of the exodus from the larger cities was sufficient to alter the traditional pattern of net migration from peripheral regions to national cores.

These studies, however, observed broad differences across Europe in the extent of this development in the 1970s. According to Fielding (1982), the move towards "counterurbanization" – identified as a negative correlation between the net migration rates and population densities of regions in a country – was most conspicuous in the Netherlands, Denmark, the UK and Switzerland. It was also evident in Sweden, Germany, Belgium and France, but Italy, Spain, Portugal, Ireland and Norway were appearing to experience continued regional concentration. Similarly, Cochrane and Vining (1988) drew a distinction between a "northwest Europe" pattern, associated with a complete reversal of migration in favour of their peripheral regions, and a "Periphery of West Europe" type where the flows from periphery to core merely diminished.

Furthermore, this development has not taken place evenly over time, but appears itself to have waned in the 1980s, if not gone into reverse. A decade ago it was confidently expected that the trend towards regional population deconcentration would become more firmly entrenched, as reflected in Fielding's (1982) prediction of a full "counterurbanization relationship" coming into being in the 1980s. It was also felt that the phenomenon would develop more strongly in southern Europe, as time went on (Hall & Hay 1980). These anticipations were, however, borne out in only isolated cases.

106

Table 7.2 Trends in urbanization and counterurbanization in the 1970s and 1980s for selected countries.

Country (number of regions)	1970s	1980s	Shift	Early 1980s	Later 1980s	Shift
Austria (16,8)	+0.38	+0.01	−	−0.25	+0.47	+
Belgium (9)	−0.36	−0.44	−	−0.49	+0.33	+
Denmark (11)	−0.79	−0.01	+	−0.04	−0.16	−
Finland (12)		+0.69	?	+0.51	+0.80	+
France (22)	−0.26	−0.36	−	−0.33	−0.31	nc
FR Germany (30,12)	−0.29		?	−0.63	−0.08	+
Ireland (9)	−0.43	−0.35	−			
Italy (13,20)	+0.12	−0.21	−	−0.16	−0.33	−
Netherlands (11)	−0.83	+0.12	+	−0.24	+0.46	+
Norway (8)	+0.21	+0.69	+			
Portugal (17)	+0.36	+0.52	+	+0.39	+0.53	+
Sweden (12,24)	−0.26	+0.35	+	+0.14	+0.53	+
Switzerland (11)	−0.49⋆		?	−0.51	−0.06	+

Notes: Data are correlation coefficients of relationship between net migration rate and population density; "Shift": + = shift towards "urbanization"; − = shift towards "counterurbanization"; ⋆data for population change (not migration); nc = no change. Also note that the correlation coefficients should be interpreted with care because their significance level depends on the number of regions.
Source: compiled from Fielding (1982, 1986, 1990), and calculations from data supplied to the author by national statistical. Reprinted from Champion (1992).

One line of evidence on these recent trends is provided by the updating of Fielding's earlier analyses, as presented in Table 7.2. This shows that more countries were experiencing "urbanization" or regional population concentration (signified by "+" coefficients) in the 1980s than the 1970s. It also shows that even more countries saw a shift in trend in that direction between the two decades (denoted by a "+" in the first "shift" column). Moreover, from the final three columns of Table 7.2, it can be seen that both these trends accelerated between the first and second halves of the 1980s.

At the same time, however, Table 7.2 also suggests considerable diversity of national experience both in the range of tendencies indicated by the correlation coefficients for any single period and by the variety of "shifts" between periods. For instance, between the 1970s and 1980s both Ireland and Italy moved strongly against the general trend, switching from regional concentration to deconcentration, and in the case of Italy this process appears to have deepened during the course of the 1980s. France also saw a strengthening of its deconcentration relationship between the two decades and seems to have maintained this pattern through the 1980s.

More detailed case studies also point to considerable complexity within countries. For the UK, for instance, studies show a very marked population recovery for London, but this has taken place alongside a large-scale growth of towns and more rural areas in southern England (Cross 1990, Champion & Congdon 1992). Similarly, the 1990 Census results for France, while indicating that the Paris agglomeration matched the national rate of population growth in the 1980s, also revealed that the highest rates of growth occurred in a broad zone stretching from the Alps through to the centre-west part of the country (Jones 1991). Important differences in settlement system change have also been noted between northwestern, central and southern parts of Italy (Coombes et al. 1989, Dematteis & Petsimeris 1989).

Drawing together the main findings of this descriptive account, it would appear to be extremely difficult – and no doubt dangerous – to use past patterns of population change on their own as a basis for anticipating future trends in the regional map of Europe. It seems clear that the 1980s experience was markedly different from that of the 1970s, with the latter itself being seen as being quite distinct from those of the previous two decades. Equally important is the fact that the levels of regional change in the 1980s were much more subdued than those of the earlier periods, with only vestigial signs of the traditional rural–urban population shifts, with rather limited evidence of the continuation of more recent counterurbanization tendencies, and with no new form of geographical patterning coming strongly to the fore. How, then, should these latest patterns be interpreted, and what significance should be attached to them in relation to future expectations?

Interpretations of the 1980s experience

Given that the regional patterns of the 1980s do not provide a clear basis for a simple extrapolation of trends in migration and population redistribution into the future, it is necessary to look behind the patterns themselves and attempt to make sense of them in terms of developments in underlying processes. This is by no means an easy task. For one thing, the marked difference between the 1970s and the 1980s makes it difficult to conceive of any long-term evolutionary trajectories. For another, the diversity of experiences exhibited by the individual countries prompts caution in any attempt at plotting out a single path for the whole of Europe.

The uncertainties involved are well reflected in the lack of consensus which currently exists between researchers concerning the significance of the events of the past decade. As this section goes on to show, probably the most widely held view is that something anomalous has been occurring at some stage since the late 1960s, but opinion is divided as to whether it is the 1970s or the 1980s that constitutes the joker in the pack. An alternative approach, which attempts to reconcile the two periods, sees the process of regional and urban change in cyclic terms, possibly superimposed on a longer-term evolution of settlement systems. Other interpretations involve the identification of two or more processes, or sets of factors, operating relatively independently and producing distinctive aggregate outcomes at different times.

THE "ANOMALY" PERSPECTIVE

The idea that something anomalous has been occurring at some stage since the late 1960s revolves around the existence of what Frey (1993) calls "period effects". This term refers to time-specific factors that temporarily interrupt the underlying trajectory of the system. For those interested in future trends in European population distribution, what is most problematic about this interpretation is that there is as yet no agreement as to whether it was the "counterurbanization decade" of the 1970s that was the anomaly or the subsequent waning of this development in the 1980s.

There would seem to be evidence available to support both these alternative views. On the one hand, several of the economic and demographic factors cited as explanations for the counterurbanization of the 1970s are now seen to have been unique to that decade. For instance, the mid-decade recession following the Arab–Israeli war and oil-price rise of 1973–4 produced a major shock to the system, including disinvestment in manufacturing activities in traditional industrial areas and the rundown of the guest-worker movement to northwest European cities. Population gains in more peripheral and rural regions can be linked directly to an increase in oil and natural resource exploitation, exacerbated by the energy crisis, as well as to a rise in expenditure on defence installations and related research establishments. This was also the time of expansion for higher education, also often away from the major existing population centres, together with the general upgrading of public-sector infrastructure and private services in rural areas.

On the other hand, the following decade also contained its share of "surprises" that can be considered equally anomalous in terms of what had happened previously. Perhaps most importantly, the 1980s were haunted by the impacts of worldwide economic recession, with setbacks early in the decade followed by deeper depression at the end of the decade. Associated with this was the slump in prices of oil, minerals and agricultural products, the restructuring of manufacturing operations designed to increase efficiency in an era of immense competition (with new "flexible" approaches replacing the traditional Fordist mode of production), and the eclipsing of Keynesian approaches to the management of the economy. Shortage of public funds, resulting from shrinkage of the tax base and – in many countries –

also from an ideological switch away from the welfare state towards greater individual responsibility and freedom of choice, led to major cutbacks in public-sector infrastructure investment, in subsidies for declining and transforming industries (with the major exception of the European Community's Common Agricultural Policy) and in spending on regional development policy. Coupled with this was a switch in government policy from schemes for urban population dispersal towards policies for regenerating inner city areas, reacting to the metropolitan problems of the previous decade and to the collapse of the baby boom that had in the 1960s and early 1970s prompted plans for new-town development projects. Also contributing to the changes of the 1980s was the warming up of the Cold War, culminating in the remarkable political events in eastern Europe at the end of the decade, with their many impacts including reduced defence spending and large-scale immigration into West Germany and Austria.

Reinforcing the idea of the 1980s being the anomalous period is the number of studies which have portrayed the 1970s experience as a logical evolution of the trends of the 1950s and 1960s, suggesting a shift over the three decades from population concentration towards progressively deeper and wider deconcentration. As shown earlier in the chapter, urbanization predominated across Europe in the 1950s, but already in the 1960s it had apparently begun to weaken. Fielding (1982) observed that in Austria and the UK the relationship between migration gains and urban status was much weaker in the 1960s than in the previous decade, while Hall & Hay (1980) found that the rings of metropolitan areas were by then growing much faster than the cores in the UK, Germany and Scandinavia. On this basis, the three decades seem to tie together as a period of progressive transition that was brought to an end by the events of the 1980s, thereby confounding the forecasts of the early 1980s.

THE "CYCLE" APPROACH

Given the fluctuations in the pace of deconcentration which have occurred over the past two decades, there have been suggestions that these are linked together in some form of cyclic relationship. One example of this approach is the concept of stages of urban development, notably the model expounded by Klaassen et al. (1981), building on the early work of Hall (1971), on England and Wales. This hypothesizes that a sequence of stages can be recognized in the development of any urban centre by reference to the relative growth of its core, ring and whole agglomeration. On this basis, a place would proceed through a "life-cycle" beginning with the "urbanization" phase of faster core growth and thus population concentration within the urban region, followed by a "suburbanization" phase of faster growth of the ring. A third phase termed "disurbanization" is distinguished by the urban

110

region as a whole experiencing decline and the core performing worse than the ring, after which the core again overtakes the ring in a period of "reurbanization", which ushers in a new cycle.

The applications of this approach need not be restricted to the study of individual urban areas, but by examining the phase reached by all urban areas in a wider (e.g. national) system, the stage-based evolution of the whole urban system can be monitored. In his pioneering work, Hall (1971) demonstrated the progressive "ageing" of the urban system in England and Wales, as generally the larger metropolitan areas started moving through the sequence during the 1950s, to be followed by smaller places. This approach has been applied subsequently to the wider European scene by Hall & Hay (1980), van den Berg et al. (1982) and Cheshire & Hay (1989). Most recently, Geyer & Kontuly (1993) have used a similar approach in propounding their concept of "differential urbanization", setting out criteria for recognizing the stage of development reached by an urban system while allowing that the urban areas within it can at a particular time be at different (but linked) stages.

The appeal of this approach lies in the fact that empirical studies of the urban-system changes which occurred during the first three decades of the postwar period appear to bear out the changes predicted by the model. In the words of Hall & Hay (1980: 87), for instance:

> In the 1950s, European population was concentrating remarkably into the metropolitan cores . . . But by the 1960s, a reversal had taken place: although metropolitan areas were still growing, they were decentralizing people from cores to rings. After 1970, this process accelerated, so that cores virtually ceased to grow.

Further evidence in support of this interpretation came from the early signs of "gentrification" in the older cores of some larger metropolitan centres during the 1970s, leading on to more significant levels of urban regeneration in the following decade.

What the "stages of urban development" argument lacks in itself is a clear theoretical structure identifying the processes involved, but other sources can provide examples of cyclic behaviour. Indeed, some elements can be seen in the discussion above on the differences between the 1970s and 1980s, with the rise and fall in prices of oil and raw materials, the expansion and subsequent rundown of various public-sector spending programmes, and the swings of government policy from urban reconstruction to regional development and back to inner-city regeneration. Cyclic patterns are also evident in the demographic patterns of recent decades, notably with the baby boom of the later 1950s and early 1960s being associated with strong suburbanization and deconcentration pressures, and being followed by a period of low fertility when not only were the pressures for suburbanization reduced but also the large baby-boom cohort started reaching the age of gravitating to the larger urban centres for higher education and employment.

It is, however, economic cycles that probably play the dominant role in any identifiable cyclic behaviour affecting migration and population redistribution, whether the causes are directly economic or the outcome of autonomous factors. Particularly well known are the relatively short-term business cycles that are associated with fluctuations in labour and housing markets and are related to employment opportunities, the availability and cost of credit, and so on. Of a rather longer periodicity are development cycles linked to the commercial property industry, notably retailing, where developers find that the profitability of one type of building or location may fall after a decade or more and therefore switch their investment to competitive alternatives before returning later to redevelop the original site. At the other extreme lies the notion of 50–60 year "long waves" or Kondratieff cycles associated with major phases of innovation and technological development, which Berry (1988, 1991) has used to account for fluctuations in the pace of US urban growth since the early 1800s.

BROADER THEORETICAL PERSPECTIVES

While the cyclic approaches can be exemplified by reference to particular factors which may fluctuate in their strength and perhaps also their nature over time, there is a third group of perspectives that provides a broader process-orientated interpretation of recent urban and regional trends. These attempt to identify groupings of factors that operate in different ways or over different timescales. These factors then combine together to produce distinctive spatial impacts for particular periods.

One perspective is that by Champion & Illeris (1990), who primarily attempt to account for the fluctuations over time in the degree of population concentration and deconcentration. Their approach recognizes the existence of three distinct sets of factors influencing the distribution of people and jobs. First, there are several forces operating over the longer term in favour of deconcentration, such as the improvement of transport and communications, the increasing preferences for owner-occupied housing, and the growth of tourism and outdoor recreation. Secondly, some factors pull towards concentration in large cities and more urbanized regions, particularly the growth of business services and corporate headquarters and other activities requiring a high level of national and international accessibility and a large supply of highly qualified manpower. A third group comprises factors that may have different geographical effects at different times depending on prevailing circumstances, operating along cyclic lines as described in the previous section. The key to applying this perspective is to acknowledge that the two sets of centrifugal and centripetal forces are both likely to vary in their relative strength over time, with an outcome which may reinforce or counter the effects of the third group of factors at any one time.

A second approach in this category is the perspective advanced by Frey (1987; see also 1993), which identifies three sets of factors in producing recent changes in the distribution of population. One of these comprises the period effects described earlier, so the principal focus here is on the other two which are seen as long-term ingredients of urban and regional change. One, termed the "population deconcentration explanation", refers to the shift from larger cities and more heavily urbanized regions to less densely populated areas and down the metropolitan/urban hierarchy, responding primarily to people's desire for lower-density housing and access to countryside and to employers' search for lower operating costs and less congested working conditions. The other is the "regional restructuring explanation", which refers to shifts in the space economy as it adjusts to new locational requirements of both production and service industries and normally results in changes in the balance between macro regions. Both these sets of processes are liable to fluctuate in their nature and strength over time, so that there may be certain periods when the two may be closely aligned in their spatial outcomes and others when they are diametrically opposed or not related at all.

Probably the most sophisticated approach to date is that of Fielding (1993a), comprising a conceptual framework for understanding the migration processes affecting South East England but one that is sufficiently general in structure to be applied in other national contexts. This model is very largely economically driven and comprises three levels according to the rapidity of change: "economic conjuncture", comprising short-term, surface-level processes like the business cycle and the housing market boom and bust phenomenon; "economic restructuring", involving medium-term, medium-level processes such as those relating to changing spatial divisions of labour; and "globalization", constituting deep-structural processes which change only extremely slowly in their character. These three levels of processes are seen as combining to produce the overall pattern of spatial changes, but according to Fielding's presentation only the middle-level processes could be responsible for the type of shifts observed in Europe over the past two decades, because the business cycles generally occupy less than a decade and have no marked effect on longer-term patterns, while the deep-structural processes evolve gradually over decades.

AN ATTEMPT AT INTEGRATION

This treatment of the separate interpretations of recent trends in population redistribution, even though it concentrates on general perspectives and does not explore some of the more specific theories such as those relating to economic restructuring and political change, presents what appears to be a bewildering variety. On the other hand, the various ideas are not mutually exclusive, but contain some common or overlapping elements. These can now be brought together to provide a more integrated perspective, which should be treated as very tentative – more a hypothesis to be tested than a statement of fact.

It is suggested that two separate processes can be held largely responsible for the swings in population redistribution trends away from and back to regional concentration observed quite widely across Europe over the past quarter of a century. One is the regional-restructuring process which, although it involves forces that are continuously unfolding, tends to provide a major recasting of the macro-region map of economic activity at relatively long intervals, principally those of the long (50–60 year) waves. The other is the population-deconcentration process, which in its strict definition refers to a general shift towards lower-density occupancy of urban space. This latter process is seen as long-established and deep-structural in nature, rooted in traditional suburbanization which still continues but also including the search by a significant number of residents and employers for smaller towns and more rural areas, which usually takes place within the same macro region but can spill over into adjacent regions if such conditions are not available more locally. Each of these two major processes comprise a larger number of specific elements, which are not always – if ever – pulling towards a single spatial outcome and which, even if they do not contain an inherent cyclic dynamic, are likely to vary in their nature and strength over relatively short periods owing to changes in prevailing conditions.

Applying this approach to recent developments in population redistribution, the early 1970s are seen as a period when these two processes were reinforcing each other. Very strong population-deconcentration forces were operating at this time, with the movement of residents and jobs down the urban hierarchy within macro-economic regions and in many cases spilling out of regions dominated by large metropolitan centres into less heavily populated regions. This coincided with the reorientation of economic activity to more peripheral regions, related not only to mineral and tourist-resource development but also to what is now seen by some as the "last fling of the Fordist mode of production", with industry actively seeking out cheaper land and labour. By contrast, the 1980s are seen as a period when the factors favourable to population deconcentration were operating much more weakly, notably as a result of the difficult economic conditions which led to cutbacks in investment in housing, manufacturing and the public sector but also attributable to demographic and other changes. At the same time, a major restructuring of the economic system was taking place, involving a fundamental reorganization of manufacturing industry and the rapid growth of high-level services particularly in the financial sector.

Looking towards the future

From what has been said in this chapter so far, it is clear that there is unlikely to be a strong consensus about trends in migration and population redistribution in the future. Besides the normal forecasting problem of not knowing for certain how the underlying factors which affect population distribution will themselves develop, the task is complicated by two additional problems. First, as now recognized by studies of past trends, the so-called "period effects" appear to have had some very significant impacts on patterns in the short-to-medium term of 5–20 years and, because of their inherent unpredictability, are likely to confound current anticipations as much as previous ones. Secondly, despite my attempt at producing an integrated interpretation of past trends in the previous section, it has to be recognized that there exist profound disagreements amongst researchers as to the validity of the alternative perspectives.

This inevitably feeds through to pose serious problems for the scenario-writers. This has recently been acknowledged very forcefully by the organizers of the Europe 2020 project which drew on the expertise of over sixty scholars from nineteen countries in an attempt to sketch out a geography of Europe's future. Having discussed trends in a range of other areas such as population size and structure, lifestyles, economy, environment, transport and communications, they found that the most controversial issues were those concerning patterns of regional development, observing that, "The analysis of the expert responses has shown that issues of regional development in Europe are extremely complex and controversial. There is little agreement about the likely future pattern of growth or about the impact or effectiveness of possible policies" (Masser et al. 1992: 105). There also seems to be a general reluctance to pronounce specifically on population redistribution prospects; for instance, the migration coverage in the Council of Europe's report on the future of Europe's population (Cliquet 1993) is restricted to international migration, even here making very little reference to its geographical impact within Europe.

What follows, therefore, is an attempt to steer through very poorly charted waters and comprises very largely a review of the various scenarios and forecasts put forward by others. It begins by examining expectations about future trends in economic restructuring and their spatial implications, taking a broad Europe-wide perspective. It goes on to look at the prospects for population redistribution along the concentration–deconcentration dimension, with the primary scale of focus here being on regional and urban patterns within countries. There then follows an examination of two sets of scenario exercises, ending with a discussion which emphasizes the major areas of uncertainty and inevitably ends on a cautionary note.

ECONOMIC RESTRUCTURING AND ITS SPATIAL IMPLICATIONS

Everyone appears to agree that the past decade has been a period of great turmoil, resulting from changing markets, new methods of business organization and major political transformation. There is, however, much less consensus over whether these changes have largely run their course or will build on themselves cumulatively. The proponents of long-wave theory believe that the main ingredients of the Fifth Kondratieff are now in place, orientated around advanced telecommunications and new industries like biotechnology, but presumably some large spatial readjustments can be expected during the 25-year growth phase of this new cycle. "Flexible specialization" has already produced severe structural upheaval, but institutional and political changes associated with the European Union and the removal of the Iron Curtain have the potential for a large-scale redrawing of the economic map of Europe. The main argument centres on whether these changes will inevitably lead to the growth of a European core region at the expense of the continent's peripheral zones.

If there was a competition for the one single aspect of Europe's future regional scenario that has captured the most attention and caused the most controversy, the prize would go to the "Blue Banana". Based on a report prepared for the French government by the Montpellier-based research centre RECLUS (1989), this term refers to the shape of a wide curved zone stretching from southeast England through Benelux, southwest Germany and Switzerland to Lombardy in northwest Italy (Fig. 7.2a). This is seen as the principal area which, over the next few years, will draw to itself even greater levels of economic activity and population than previously, benefiting from the introduction of the Single European Market at the end of 1992, the further steps towards European economic and political integration presaged by the Maastricht agreement and by the enlargement of the European Union, and the plans for the further development of TGV-type rail systems and telecommunications infrastructure. Places lying within this broad zone, whether large or small but particularly those at nodal points in the evolving system, are viewed as having a major advantage over places outside this zone, with their growth leading to a further widening of economic disparities between core and periphery in Europe.

This bald prognostication has been criticized from several quarters, not least by those who view it as far too simplistic (see Masser et al. 1992: 92–107). One point is that past experience, for instance that of the "Golden Triangle" based on London, Paris and the Ruhr, suggests that not all places within a favoured zone do indeed prosper and that location outside it is not necessarily a serious disadvantage. Secondly, noting the strong economic growth achieved by northern Italy, southern France and more recently by northeast Spain, some pundits reckon that a

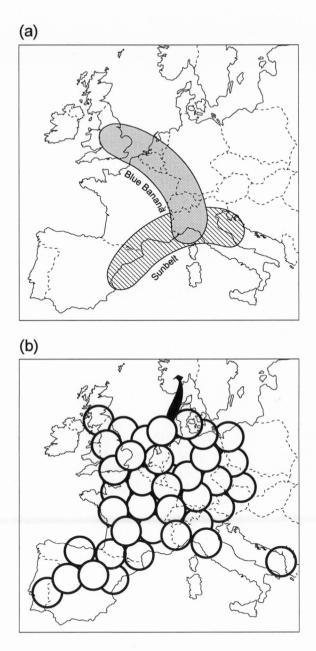

Figure 7.2 Alternative fruity images of Europe's future pattern of economic development: (a) bananas; (b) grapes. *Source:* after RECLUS (1989), Kunzmann & Wegener (1991).

Mediterranean "Sunbelt" is emerging as a second major European growth zone (Fig. 7.2a). Thirdly, while few experts involved in the Europe 2020 project hold out any great hopes for strong economic growth in the "Atlantic Arc", a fair number recognize that certain forces will counter the tendency towards ever greater concentration; notably, the costs of greater congestion and environmental damage in the fast-growth areas, the greater freedom of locational choice allowed by improved transport and communications, and the necessity for effective government intervention to prevent the disintegration of the European Union that increasing national and regional disparities would foster.

Even more fundamentally, however, the whole concept of European core zones structured around physical transport links has been challenged by research which concludes that the economic fortunes of individual places have much more to do with their inherent characteristics and the way in which they position themselves with respect to the global economy and international capital than with their particular geographical location. This approach has its roots in the "localities" research tradition (see, for instance, Cooke 1989), which noted how often neighbouring places have developed in very different ways and traced the cause to various factors relating to the quality of entrepreneurship and to their diverse complexions of inherited advantages and constraints. This perspective sees Europe as made up of many individual, largely urban-centred regions which compete for jobs, people and capital investment, sometimes in cooperation with their neighbours but as often as not being in competition with them while perhaps being linked to other similar regions elsewhere through such networks as the Eurocities Group, POLIS and the Commission des Villes. Rendered into the crude fruity analogies that were very popular at the beginning of the decade, this perspective is portrayed as the "European grape", as shown in Figure 7.2b (Kunzmann & Wegener 1991), while Illeris (1992) speaks of a regional "mosaic" (see also Champion & Illeris 1990).

Following this perspective, it is now common for both analytical and forecasting studies to divide up Europe not in terms of broad geographical zones but into types of regions. For example, Illeris (1992) feels that the main elements of the new map of population redistribution in western Europe are best thought of in terms of up to six types of regions: national capitals, old industrial and mining regions, other large cities, traditional agricultural/rural periphery regions, dynamic regions with small and medium-sized towns, and tourist regions. Although Illeris finds considerable convergence in their population growth rates between the 1960s and 1980s, this is seen as a period effect and a reopening of the differentials is expected as a result of economic recovery and the greater regional competition resulting from the Single European Market. This approach is developed in greater functional detail by Kunzmann & Wegener (1991). They point to growing evidence that certain modern industries prefer certain types of cities:

Worldwide economic concentration, agglomeration economies and information and communication synergies reinforce each other and favour polarization in a few cities at the expense of others. These cities then specialize
their infrastructure and services to attract additional economic activities of
the same kind, until the city has a 'label' or image for a particular mix of
activities. (Kunzmann & Wegener 1991: 35–6).

With this in mind, they present the functional categorization of European cities
shown in Table 7.3, listing dominant characteristics and exemplar cities for each of
the eleven types, and then go on to assess the current situation and future prospects
of each type. According to them, for instance, the global cities of London and Paris
have a self-reinforcing mechanism which will attract further new activities, the
growing high-tech/services cities can expect continued strong growth despite cutbacks in defence expenditure, and border and gateway cities will develop very
strongly along the new economic corridors. By contrast, the prospects for declining
industrial and port cities are felt to be generally poor, while company towns, mono-
functional satellites and small towns in more rural areas present a potentially very
varied picture, being dependent on a narrow sector or even a single firm.

In similar vein, Parkinson (1991) has put forward a three-dimensional framework for classifying Europe's cities. One dimension relates to the spatial markets in
which cities will be operating, distinguishing the three primary levels of regional,
European and global. The second, termed "urban success", is measured in terms of
recent economic performance, environmental and cultural attractiveness, and level
of social cohesion. The third takes account of the regional context of each city, dif-
ferentiating between growing, adjusting and developing regions. On the other
hand, while classifying several exemplar cities on the basis of these three criteria,
Parkinson acknowledges that there are practical problems in using this approach to
predict a ranking of individual cities in terms of growth potential, notably the difficulty of attaching weights to each of the three dimensions and the current lack of
knowledge about the internal political and entrepreneurial dynamics of cities.

URBAN CONCENTRATION AND DECONCENTRATION TENDENCIES

In the interpretation of recent trends outlined earlier, the "population deconcentration" explanation is seen as operating alongside economic restructuring as a second major dimension of population redistribution and migration. At the same time
as any locality's growth performance is being affected by its inherited economic base
and its ability to attract new investment, it will also be influenced by any general
tendency either towards population dispersal out of major metropolitan centres and

Table 7.3 Functional types of cities.

City type	Characteristics	Examples
Global cities	Accumulation of financial, economic, political and cultural headquarters of global importance	London Paris
Growing high-tech/services cities	Modern industrial base, national centre of R & D, production-oriented services of international importance	Bristol Reading München
Declining industrial cities	Traditional (monostructured) industrial base, obsolete physical infrastructure, structural unemployment	Metz Oberhausen Mons Sheffield
Port cities	Declining ship building and ship repair industries, environmental legacies, in the south burdened by additional gateway functions.	Liverpool Genova Marseille
Growing cities without modern industrialisation	Large informal economy and marginalised underclass, uncontrolled development and deteriorating environment.	Palermo Thessaloniki Napoli
Company towns	Local economy depending to a high degree on a single corporation	Leverkusen Eindhoven
New towns	New self-contained cities with overspill population in the hinterland of large urban agglomerations	Milton Keynes Evry
Monofunctional satellites	New urban schemes within large agglomerations with focus on one function only (e.g. technopole, airport)	Sophia-Antipolis Roissy
Small towns, rural centres urban belts	Smaller cities and semi-urbanised areas in rural regions, along coasts or transport corridors with weak economic potential	All over Europe
Tourism and culture cities	Local economic base depending on international tourism and cultural events of European importance	Salzburg Venezia
Border and gateway cities	Hinterland divided by national border, gateways for economic migrants and political refugees	Aachen Basel

Source: Kunzmann and Wegener (1991).

more heavily urbanized regions or towards greater concentration. Although such shifts down or up the urban hierarchy could result at least in part from the aggregate effects of economic restructuring, as seemed to be the case with the counterurbanization of the 1970s, here it is treated as a separate phenomenon arising from a distinct set of factors which operate at intra-national levels and primarily within macro regions. Following the lead of Champion & Illeris (1990), it can perhaps be visualized best as a tug-of-war between the forces of concentration and deconcentration, with both operating simultaneously but varying in their relative strengths.

Age-specific migration provides one clear example of this tug-of-war process, revolving principally around the attractive and repulsive roles of different elements of the urban system for various population subgroups. Retirement migration, for instance, very largely takes the form of population movement away from larger metropolitan centres and more densely settled areas and down the urban and regional hierarchy, taking advantage of the lower house prices and more congenial environments of smaller towns and more rural areas, with relatively few 50 to 70 year olds moving in the opposite direction. The suburbanization process has traditionally been powered very largely by couples reaching the family-building stage of their life-cycle and seeking affordable houses with gardens and a safe and pleasant environment within commuting distance of work. By contrast, young adults tend to be drawn to the larger cities to take advantage of their higher education facilities, better job-seeking networks, more varied social opportunities, cheaper rented housing, and so on.

The outcome of the "contest" between this particular set of centrifugal and centripetal forces depends primarily on the sizes of the respective age groups in the population and how these are altering over time. The number reaching retirement age was growing strongly in the 1960s and 1970s in all European countries, but has subsequently shrunk somewhat as a result of the smaller size of the birth cohorts of the 1920s, presumably reducing the numbers available for leaving the large cities. Meanwhile, the level of family rearing activity declined in most countries after the early 1970s, weakening the demand for new housing in the suburbs or new satellite towns, and the 1980s saw the "coming of age" of the 1960s baby-boomers and thus an increased potential for young-adult migration to metropolitan centres. Changes in age composition can thus be seen to parallel rather closely the general European trend of strong population deconcentration in the 1970s and a revival of concentration tendencies in the 1980s.

Forecasting the future balance of metropolitan push and pull is a fairly straightforward exercise as far as this one factor is concerned, because age-based population projections are available for all countries. The latter show that in the European Economic Area the number of people reaching the age for retirement-type moves will remain at relatively low levels until the end of the century but will then increase steadily in response to higher post-1945 fertility and reach a peak in the 2020s. On the other hand, the 1960s baby-boom cohorts have now passed through their young-adult phase and reached family-building age. All other things being equal, therefore, the demand for large-city residence should generally be weaker now than in the 1980s and the centrifugal forces associated with family-rearing should be running at a higher level.

Unfortunately for forecasting, however, other things are rarely "equal". There are various factors which can prompt people in the various age groups to modify their behaviour, some of a short-term nature but others more deeply structural.

121

Retirement migration propensities, for instance, vary according to the state of metropolitan housing markets, obviously being highest when prices are high and turnover is rapid. In some countries where there is a post-war history of strong net rural–urban migration, such as France, much retirement migration has involved return moves of the original migrants to homes that they have inherited from their families, but this particular route is becoming less common as more people are city-born and more rural housing is sold to others.

Also impinging on age-specific migration behaviour are trends in household composition, notably the changes in household-formation behaviour which have taken place since the mid-1960s. Whether or not one subscribes to the notion of the "second demographic transition" (van de Kaa 1987), there is no doubt that the past thirty years have witnessed the rise of a variety of non-traditional living arrangements and life-styles. These are most readily reflected in statistics on trends in household size and composition, in divorce and cohabitation, and in the level and incidence of child-bearing (Hopflinger 1991, Champion 1992, see also Ch. 3 of this book). For example, in much of Europe average household size is now below 3.0 persons, the proportion of households that contain only one person is over 25 per cent, at least one in ten families are headed by a lone parent, the divorce rate is running at around a third of marriages, the total fertility rate is well below replacement level and more women are starting their family later in life. Such dramatic changes cannot but affect migration, but as yet very little is known for certain about the geographical impacts of these changes. Problematically, two contradictory outcomes can be expected: the increase in number of smaller, less wealthy households resulting from these changes would normally lead to greater overall preference for the smaller, cheaper housing units that have traditionally catered for young adults in the inner areas of the larger cities, yet the sheer absolute volume of extra households could overwhelm the housing capacity of these cities and prompt a higher level of overspill by these or other households.

In addition to the effects of these compositional changes, it must be recognized that the migration preferences of any given population subgroup can alter over time. After all, it must be remembered that the early research on "counterurbanization" identified changes in residential preferences as the key factor in instigating the quest for the new open spaces (Berry 1976) and that "fashion" can be as important in the search for a house and living environment as in the other aspects of life. Once again, retirement migration provides a clear example of this, with the most popular destinations switching from the seaside and spa towns of the 1950s to the countryside areas of the 1960s and 1970s and to the Mediterranean sunbelt zones of the 1980s. For younger people the growing preference for owning rather than renting has acted as a very powerful agent favouring population deconcentration, but the recent difficulties in property markets, together with the demographic changes just mentioned, seem to have dampened this tendency, at least for the present.

A key part of the preferences debate revolves around the increasing attention being given to "quality of life" considerations in residential decision-making (see Findlay & Rogerson 1993 for a review). Although "quality of life" can be defined broadly so as to include access to employment and cost of living, the term is used most often to refer to the physical and social aspects of living standards, with most weight being given to low crime, congestion and pollution levels and good climate, schools, shops, and leisure and health-care facilities. Over the decades these features have become increasingly associated with suburb and countryside, as conditions in the older urban cores have deteriorated and service provision has extended progressively into more rural areas. Recent events, however, seem to be clipping the wings of this force for deconcentration, with the greater penetration of negative features into the countryside (including congestion and crime), with the withdrawal of both private and public-sector services from rural areas through rationalization, and with older urban areas beginning to benefit from the regeneration policies. Even so, it remains to be seen how far this rebalancing of the relative merits and disadvantages of city and suburb/countryside will proceed; in short, whether there is a natural "cycle" in the life of cities and whether a clear "re-urbanization" phase is being launched.

The possibilities of this happening do not, of course, depend entirely on people's preferences but also on decision-making by employers. Although the latter was seen above as being particularly relevant for the distribution of economic activity between macro regions, it also has some influence on population patterns within regions. Once again, however, recent events have not flowed in one single direction; for instance, with the growth of business services not only reinforcing the position of the largest cities but also spawning a large volume of "back-office" jobs in suburban areas and more distant towns alongside decentralized factories and shopping complexes. In terms of future residential implications, much depends on how far workers are prepared to commute and also on the extent to which "tele-commuting" (working with computer, phone and fax at a remote location at home or teleport nearby) can substitute for daily physical presence in the main office.

In this, as in other aspects of the "tug of war" between centripetal and centrifugal forces, there is a complex mixture of choice and constraint. In relation to worker/employer interrelationships, the traditional view is that the employer makes the key decisions and the employee operates within the framework that derives from these. Nevertheless, the availability of certain types of labour has at particular times been able to exercise considerable influence over firms' location decisions; for instance, in prompting relocation to low-wage regions and in attracting research activities to localities highly prized for their natural beauty or social kudos. During the 1980s much policy discussion in Europe was focused on the prospect of a major increase in labour power caused by the marked decline in numbers of school-leavers, one aspect of the "demographic time bomb" set ticking by the 1970s baby-bust

– though, so far, this has largely been defused by economic recession (Green and Owen 1993 and see Ch. 4).

Finally, much of the context for decisions taken by both individuals and firms is provided by planning authorities of one sort or another and by the policies which they have pursued for modifying new development patterns. As noted above, "counterurbanization" has been attributed in part to the role of the public sector in its dispersal schemes and in those sectoral investment programmes that helped to reduce locational constraints. With the disengagement of governments from spatial policy and general spending being seen as one of the causes of lower rates of deconcentration in the 1980s, the question arises as to whether this, too, is subject to cyclical behaviour and thus whether the next few years will bring an upturn in the role of the state. Recently expressed concerns over environmental and conservation issues (e.g. energy usage, global warming, environmental degradation) suggest that this may well be the case, although the precise effect of any action on deconcentration trends is by no means clear; witness the controversy over the most energy-efficient form of settlement pattern (Masser et al. 1992).

To conclude, at the local scale the decentralization tendency is well established, but in general terms it has proceeded further for larger cities than smaller ones and in the north of Europe than in the south. Even so, in most countries the level of regional and urban population concentration remains high, suggesting great potential for further dispersal if the conditions are right. What is so problematic for forecasting, however, is that the process is the subject of a mixture of countervailing pressures which vary in their nature and intensity over time. Proponents of urban life-cycle theories would no doubt anticipate an eventual return to stronger deconcentration after the current phase of large-city recovery, but the proof is lacking and the timing is difficult to gauge.

"WHAT IF?" SCENARIOS

Neither the economic restructuring perspective nor the population deconcentration approach would seem to provide the basis for firm guidance on future patterns of population distribution; and, moreover, there is no obvious way of producing a synthesis between the two. A common response to this type of problem is to predict what would happen if a particular set of circumstances were to prevail, i.e. in this case, by making assumptions about the future nature of population-change patterns or about the factors which are believed to determine them. This final section looks at two examples that take relatively extreme assumptions and thus provide some indication of the limits within which the actual events are likely to lie. Having said this, however, the concluding discussion examines some potential "surprises" which could significantly reduce the relevance of these scenarios.

The first of the examples, by Rees et al. (1992), comprises quantitative projections designed to explore the impact of particular sets of assumptions about migration, using the Level 1 regions of the then twelve European Union member states. Altogether, four scenarios are developed, covering the period from 1990 to 2020. The first assumes zero interregional migration and provides a marker against which intra-Union migration impacts can be measured, in that it looks only at the effects of births, deaths and migration between the twelve and the rest of the world. The second envisages the continuation of mid-to-late 1980s migration rates, applied to the changing regional populations.

It is, however, their two "regional scenarios" that are of particular interest here, since these adopt alternative approaches to the nature of regional migration. A "growth regions" scenario assumes a direct relationship with existing regional GDP per capita levels and attractiveness to net migration, so that the richer regions gain more migrants and the poorer ones lose more. The other scenario, termed "counterurbanization/urbanization", is based around the relationship between net migration change and regional population density, but adopts different arrangements for southern Europe compared to the rest of the Union: a positive (i.e. concentration) relationship for the former as opposed to a negative (i.e. deconcentration) one for the latter. This final scenario essentially mirrors the experience of the 1970s, while the "growth regions" scenario represents an exaggerated version of the underlying trends of the 1980s.

Not surprisingly, these two alternatives to the constant-migration scenario produce some large differences in overall population change rates, both in terms of general patterns and for specific regions. Under the growth-regions approach, the rich capital cities or core regions increase their populations at the expense of poorer, peripheral regions, with the population of Brussels rising by 53 per cent over the 30 years to 2020 and Hamburg by 28 per cent, much higher than the 14 per cent gain and 11 per cent loss projected for these places on the basis of constant migration rates. Under the counterurbanization/urbanization scenario, the results for southern Europe are broadly similar to those from the growth-regions scenario because both approaches have the largest gains occurring in the rich, more densely populated urban regions. For northern Europe, by contrast, the results are generally the opposite of the growth-regions approach, with the migration shift away from the wealthy urban concentrations: to the extent that Brussels is projected to experience a 14 per cent population decline and Hamburg's size is halved.

Perhaps the most significant lesson to be learnt from this example is that migration is potentially a very major force for regional change in Europe. On the other hand, looking back over the past 30 years makes one realize how extreme the two scenarios are, not so much because of the specific migration rates adopted but because of the "purity" of each scenario and the assumption that a particular pattern of migration will be in place continuously for three decades. To go beyond this experimental exercise would require both the development of a series of hybrid scenarios which combined weighted elements of each, and then the setting of the start and duration of the various hybrids. The timing in the model is as crucial as the other aspects because of the way in which the regional population bases are modified incrementally by each annual round of the projection process.

By contrast, the other scenario example is purely qualitative, but adds a further dimension in that it attempts to visualize the impact of fundamentally different political stances. In their work (Masser et al. 1992) the Europe 2020 group distinguish three basic alternatives. A "growth scenario" assumes that all policies emphasize economic growth as the primary objective and would very likely also be a high-tech and market-economy scenario with a minimum of state intervention, reflecting the political ideals of many current conservative governments in Europe. An "equity scenario" assumes policies that primarily try to reduce inequalities in society both in terms of social and spatial disparities, thus being associated with the typical policy-making of social democratic governments. Lastly, an "environment scenario" stresses quality of life and conservation, involving a restrained use of technology and some control over economic activity and conforming to the views of Europe's green parties.

In their pure forms, each of these three scenarios are seen to pervade every aspect of human behaviour and societal activity, including decisions on fertility, life-styles, economy, transport and environment, and thus give rise both directly and indirectly to distinctive geographical outcomes. Thus, the growth scenario is predicted to lead to the concentration of growth in a megalopolis of 80 million people stretching from London to Milan, with recent tendencies being reinforced by the incorporation of the remainder of Europe into the Single Market and TGV-rail network and with the continent's more peripheral regions sinking into economic decline and depopulation. This scenario is also associated with the virtual disappearance of countryside around the rapidly expanding agglomerations, and within them the emergence of a clear three-fold division of space between their high-status international city cores, middle-class suburbs and underclass ghettoes.

By contrast, the equity scenario envisages a strong decentralization policy with a major enhancement of both national and supranational powers to redirect economic activities towards peripheral areas. Investment programmes such as a Technopolis Network, a Remote Areas Highway Programme and a Regional Airports Scheme would be coupled with strict land-use controls in urban areas, tax incentives for location in non-metropolitan regions and flat rates for long-distance telecommunications services. It is suggested that these policies would be successful in producing a relatively balanced choice between urban, suburban and rural areas for both residents and firms.

Lastly, the geographical effects of the rigorous policies adopted to force industry and consumers to act in an environment–conscious manner are seen to be particularly far-reaching by comparison with previous experience. Under this scenario, some regional activities such as the potash and lignite industries in eastern Europe are forced to close down completely, while others have to reorganize their production fundamentally to meet emission standards. As transport is forced to become cleaner and therefore becomes more expensive, the effect is to limit regional economic decentralization and necessitate the introduction of support measures for peripheral regions and agricultural areas. It would also encourage public transport and the fuller integration of residence and workplace at more local scales.

As with the first scenario exercise outlined above, this second example again provides some extreme "what if?" alternatives, but in relating each of the three scenarios to a political ideology, it goes further by suggesting that there are real choices to be made. Indeed, the Europe 2020 project made use of a panel of 60 experts to assess both the probability and also the desirability of the three. Not surprisingly, there was a major discrepancy between the two sets of judgements, with the growth scenario being seen as the most likely outcome but with opinion on the most desirable being fairly evenly split between environment and equity scenarios. The views on desirability are used to sketch out a target scenario, and it is tentatively suggested that attitudes will shift temporarily towards the growth scenario but in due course swing around more in favour of the equity approach and subsequently favour the adoption of some of the more radical policies associated with the environment scenario (Masser et al. 1992).

Both these scenario exercises are valuable for several reasons. First, neither is confident enough to put forward a definitive prediction of the future map of Europe, confirming that no consensus exists on this topic. Secondly, in sketching out some scenarios that are reckoned to be extreme or pure forms of potential directions of change, they plot out the limits within which the future course of events is considered to be located. Thirdly, in both cases but most explicitly in the case of the Europe 2020 project, the scenarios are seen to provide the basis for conscious choice about the future. In that sense, anyone who wants to anticipate trends in population distribution has to make some assumptions about the way in which policy attitudes will develop over time.

In this context, however, two important questions need to be raised. How effective will such policies be? This turns not only on the efficacy of state intervention in general, but on the degree of concordance between different layers of government: local, regional, national and supranational. Secondly, will there be any major "surprises" that will render current scenarios largely irrelevant? The answer is bound to be "Yes!", if the anomalies of the past two decades are anything to go by. Indeed, the recent political upheavals in eastern Europe already provide an example of an event which had scarcely been imagined two years before, but immediately began to transform the geography of the continent. The longer-term repercussions, including the military conflicts in parts of former Yugoslavia and the Soviet Union,

127

are now unfolding but remain difficult to predict. Also to be flagged as a key area of uncertainty over the next few years is the level and nature of migration between Europe and the rest of the world, itself only partially amenable to policy influence. Clearly, the possibilities are legion!

A final word

Brave this chapter has not been! It has described recent trends in migration and population distribution and set them in their longer-term context, and it has attempted an interpretation of these developments in the light of the available literature. But, bearing in mind the adage that "those who know enough to forecast migration know better than to try", it has shied away from the challenge of predicting the future course of population redistribution and has instead attempted to sketch a framework to guide those who are interested in taking this further.

It should be no surprise that there is no clear consensus about future patterns of population distribution at regional and urban scales across Europe. Even now, the patterns of change during the 1980s are not well documented, let alone properly understood, but enough is known to recognize that each decade brings fresh developments. In relation to US experience, Frey & Speare (1992) questioned whether the events of the 1980s could best be characterized as "back to the past", "no turning back" or "back to the future" and concluded broadly in favour of the last interpretation, identifying a return to urbanization but in new directions (see also Frey 1993). This probably applies even more forcefully to the European scene, with its dense and comparatively slow-changing urban networks and some evidence of cyclic behaviour on the one hand but with its recent record of rapid economic, sociodemographic and political transformation on the other.

Moreover, there remain considerable barriers to the improvement of our understanding, notably of a statistical nature. Unlike the USA, where one single organization is responsible for both the decennial Census and annual population monitoring and implements standardized procedures across an area somewhat larger than the European Economic Area, research on Europe has to deal with over 30 different data-collection systems and agencies. The European Commission has been attempting to achieve statistical harmonization for at least twenty years, but has so far made very little headway on local population and migration datasets, even for the original six signatories of the Treaty of Rome (Poulain et al. 1991). Meanwhile, the Council of Europe continues to expand its annual reporting of demographic developments, but has yet to include sections on internal migration and subnational population trends (Council of Europe 1994). The importance of regional and urban population trends as a policy issue is not yet matched by the ability to monitor and analyze them, at least at a pan-European scale.

Acknowledgements

I am extremely grateful to the Council of Europe for permission to publish Table 7.1 and Figure 7.1, and to Klaus Kunzmann, Michael Wegener and IR PUD for permission to publish Table 7.3 and Figure 7.2(b).

CHAPTER 8

The future of skill exchanges within the European Union

ALLAN M. FINDLAY

Introduction

International migration between the states of the European Union has never been easier, yet there has been no substantial increase in long-term skill exchanges over recent years. The provisions made under Article 52 of the Treaty of Rome (which took effect in 1968) meant that citizens of any member country had the right to seek and take up work and residence in another member state. This position was strengthened by the 1985 Schengen agreement which aimed at the removal of border controls between the signatories (nine EU states had signed by 1993). Despite these measures, and the increased internationalization of West European economies, it would appear that international migration between EU states actually declined during the 1980s. Strong forces continue to limit the scale of permanent skill exchange. This chapter examines the specificity of the West European context in searching both to account for this paradoxical situation, and to consider whether western Europe will continue to diverge from other major advanced trading areas in its migration experience.

Following a brief literature review, the chapter considers the limited secondary data available on international migration between European Union states. Attention is then focused on the insights that may be gained from recent survey results concerning the organization and behaviour of skilled international migration within the EU.

Business travellers, expatriates, impatriates and the internationalization of production

The internationalization of capital has been a powerful influence in remoulding the geography of global production in the 1980s and 1990s. There have been few places

130

where this has been so evident as in the core economies of western Europe (Dicken 1992). Furthermore, there is well documented evidence that as trade between countries grows so too does business travel. Ford (1992), for example, has shown that a statistically very significant correlation ($r = +0.91$) exists between the gross number of business passengers between the UK and a range of destinations and the gross value of trade with these countries. In a similar vein, McPherson (1993) has shown that trends over time in UK trade and in the number of international business passengers are strongly correlated for the period 1982–90. It would therefore appear that internationalization and economic integration are strongly linked to growth in international population movements.

Extending the analysis from business travel patterns to skilled labour migration, Findlay (1988: 408) has argued that the "hierarchically organised global system of production has reduced the need and the opportunities for settler emigration and has promoted new forms of international labour transfers". In particular, it has been suggested by Findlay & Gould (1989: 4) that the global shift of investment by large employers has "necessitated the transfer of an increasing number of managerial and technical staff to supervise operations resulting from foreign investments". These comments were made specifically with reference to research relating to skilled migration from core areas within the world economy towards the newly industrializing countries and to other destinations in the developing world (Salt and Findlay 1989).

Beaverstock (1990, 1991) has extended research on skilled transients to moves between core economies, affirming from his study of the internationalization of financial services that this has been associated with the growth of an international labour market for highly skilled staff. Salt (1988) and Findlay (1993) have noted that these skill flows between core economies take the form of "skill exchanges" rather than of a "brain drain". This results from the fact that skill moves between countries are organized by large companies in order to service the needs of their internal labour markets, with the key skill transfer often being more to do with a pre-existing managerial knowledge of the firm's operations rather than with a specific professional or technical matter. The skill exchange may well be specific to the company even although it involves international transfers of personnel. To quote from Ford:

Large organisations require individuals able to offer more general management skills across many different divisions or locations. Often these skills are more specific to the organisation than they are to any one task or responsibility . . . Individuals can find their skills applicable within the internal labour market of their employing organisation but relatively unsaleable outside. (Ford 1992: 31)

131

Applying some of these concepts to the West European context, Salt (1993: 22) concludes that migration of highly qualified staff should continue to rise as companies internationalize their business operations and as organizations become more complex and increase in size. Deroure (1992: 22), writing on a similar theme, makes the distinction between the transfer of international staff from a company's location of origin to a foreign posting, and the quite separate process of hiring foreign staff for subsequent transfer to the company's country of origin as part of an attempt to internationalize the upper echelons of companies. Deroure coins the term "impatriates" to describe this latter process. Equally significant is the discovery by Boyle et al. (1994) of the hiring of significant numbers of "expatriates" on a local contract basis. This arises partly from the ability of European firms to tap, in many of Europe's largest cities, a local labour market of resident foreign skilled staff. This process avoids the need to pay for expensive international staff transfers as well as allowing companies to economize by employing foreign staff on local rather than expatriate contracts.

Several issues arise from this brief and selective literature review which demand attention in the remainder of this chapter. First and foremost is the question of whether there is empirical evidence of a growth in skilled migration between European Union states. Secondly, it is clearly important to resolve whether increased economic linkage is likely to lead to increased migration between EU states towards the year 2000. Thirdly, it becomes apparent from the literature that there is a need to consider more carefully the different categories of skilled migrants that may be involved in European exchanges, and on the basis of a better typology of skill exchange, to consider the underlying processes which will be responsible for future developments in skill mobility.

Growth or decline of skilled migration in the EU in the 1980s and 1990s?

It is extremely difficult to identify migration trends across western Europe because of the differences in national definitions and data sources on stocks and flows of foreign population. Arguably the most reliable source of comparable data is provided by the OECD's SOPEMI system of annual reports on migration trends (Salt 1987). Analysis of these leads to the overall conclusion that during the 1980s foreign population in western Europe increased (King 1993c). By 1991 there were around 13 million foreigners legally resident in the EU, eight million of whom were non-EU nationals. Despite a decade when the migration policies of EU states were set against further immigration from outside the community, it therefore appears that immigration continued unabated in terms of the numbers of people involved.

Although the annual OECD report on international migration does not include detailed tables for each EU country, some indication of intra-EU transfers is evident. This information has been re-tabulated. The most striking feature of Table 8.1 is that it shows a decline in the stock of EU nationals in all but one of the countries for which data are available. Thus, in contrast to immigration to the EU, intra-EU moves declined during a decade when national policy-makers were moving towards ever freer transfer of people between the countries of western Europe. Rather than merely concluding that migration policies are ineffective whether they operate to encourage or discourage international population moves, it is pertinent to examine Table 8.1 in more detail.

Table 8.1   International migration between selected EU states in the 1980s.

Host country	Total stock at end of period	From all EU states: change in period		Origin country: change in thousands during period				
	(000s)	(000s)	(%)	France	Germany	Italy	Spain	Portugal
Germany (1982–9)	1325	−168.2	−11.3	+11.4	–	−53.3	−38.8	−21.4
France (1982–90)	1308	−285.9	−17.9	–	nd	−86.6	−111.2	−121.7
Belgium (1982–90)	550	−38.8	−6.6	−9.4	+0.3	−31.9	−5.2	+6.1
Netherlands (1982–90)	168	−7.0	−4.0	nd	0.0	−4.5	−5.4	+0.3
UK (1984–90)	889	+198.0	+28.2	+15.0	+8.0	−8.0	−1.0	nd

nd = no data.
Source: OECD, (1992).

It is clear that the main reason for the decline in the stock of EU workers and their dependants in Germany and France during the 1980s was the return of Italian, Spanish and Portuguese guest-worker families to their country of origin. This pattern of return flows may be explained as part of the wider industrial restructuring that was taking place in the 1970s and 1980s.

Industrial restructuring is often analyzed in relation to the concepts of regional sectoral specialization and the new international division of labour (Fröbel et al. 1980, Fielding 1993b). In the 1950s and early 1960s the regional sectoral specialization of production of goods in high demand attracted migrants to the core cities and regions of the countries of northwest Europe. The migrants came from regions specializing in the production of those goods for which demand was stagnant or declining including the countries of the European periphery. By the 1970s and 1980s the new international division of labour was producing industrial decline in the core regions of western Europe. At the same time the redistribution of industrial

133

investment towards lower wage economies was improving prospects in the peripheral regions of Europe for migrant workers wishing to return (Cavaco 1993, King & Rybaczuk 1993). Thus, as de-industrialization came to typify the core regions and cities, so return migration from core to periphery began to take place.

The most obvious exception to the pattern which has been described is the case of the United Kingdom which, as Table 8.1 indicates, saw its stock of EU nationals increase by almost 200000 during the latter half of the 1980s. This trend was in large part produced by the continued inflow of Irish migrants (Shuttleworth 1993) who accounted for more than three-quarters of the growth in EU migrant stocks during this period.

The new international division of labour may have had a second and perhaps less noticeable effect on migration between EU countries in the 1980s. Table 8.1 indicates that some inter-country flows increased between 1982 and 1990. For example, the stock of French migrants in Germany and the UK rose, as did the number of Germans in Belgium and Britain. Although the absolute increases were quite small, in relative terms they were much more significant (a 65 per cent increase in the number of French and a 24 per cent increase in the number of Germans in the UK). Those accepting the principle of the new international division of labour as the explanation of such trends (Fielding 1993b) would suggest that it reflects the emergent functions of core regions and cities of northwest Europe as headquarters locations and foci of the research and development activities of large international companies. As a result they have become locations of immigration for highly qualified labour moving within the internal labour market structures of large companies.

Ironically there are more secondary data available to show that highly qualified staff move to Europe from other core regions of the global economy (such as from Japan and the USA to the UK), than there is of skill flows connecting the core regions of western Europe. This is the case simply because of the freedom of movement offered to professional and managerial workers within the EU, a situation which contrasts with the restrictions placed on skilled immigrants entering the EU from elsewhere. For example, Japanese and American staff working for large corporations in the UK have to obtain a work permit. Britain's long-term work permit statistics show that in 1990, 40 per cent went to Japanese and American immigrants, most of whom were professional and managerial staff and many of whom were granted permits as persons making intra-company transfers (Findlay 1992, Salt & Ford 1993). Similar patterns are found in Australia, whose migration statistics for those on long-term business visits (over 12 months) show that Japan and USA account for 54 per cent of such moves.

There is therefore substantial evidence that small, but significant, high-level skill flows are taking place between the core economies of the world, organized through the internal labour markets of large companies. As far as moves between the core cities of Europe are concerned, the lack of reliable detailed secondary data means that it is only possible to infer that such flows are taking place. The positive trend in migration between France, Germany and the UK shown in Table 8.1 is compatible with the inference that European companies are operating in the same fashion as their competitors from other parts of the globe, but as will be shown below, such an inference is not entirely supported by detailed survey research.

This section commenced by asking whether secondary data pointed to an increase or a reduction of skilled international migration in the EU. It is concluded from Table 8.1 that international exchanges between EU states during the 1980s involved a decline of traditional guest-worker migration as well as a modest growth of skill flows between the core economies of northwest Europe. The data do not, however, permit one to reach a conclusion as to whether such flows were directly associated with the growth of large companies and trends towards internationalization or were brought about by other factors.

International migration, trade flows and international investment

The statistical correlation between international business travel and trade flows has already been discussed above (Ford 1992, McPherson 1993). Salt & Ford (1993) have argued that the reasons for business travel vary with the length of trip, with short trips being used for fact-finding, conferences and limited purpose meetings; visits of a week to five weeks being associated with sales activities, troubleshooting and certain forms of training; and longer trips being reserved for technical assignments and expatriate-leave relief. As trade and economic integration between countries increases it seems logical that trips of all these kinds will increase in number. McPherson (1993) goes further in his analysis by identifying whether business trips are essentially concerned with linkages between different branches of a firm, or between different firms. Using the crude distinction between manufacturing and service activities, his survey showed that business travellers representing manufacturing organizations were likely to undertake more international trips than those from the service sector as well as being more likely to travel between different branches of the same corporation. Within manufacturing, electronics and electrical engineering was the sector with the highest propensity for frequent business travel.

Despite the relationships described above, there is no necessary link between the economic integration of two or more economies and international migration between states. Since international migration, or expatriation, involves a greater permanence of relocation on the part of the employee, greater cost on the part of the employer and the demand for a different skill function in terms of the organization's purpose for supporting a longer-term skill transfer, there is good reason to

135

ask what logic exists to endorse the expectation that a growth in international migration will follow economic integration.

A recent OECD study (1992) has provided interesting data sets relating immigration to trade for the years 1975–90 (Maghreb to France; Mexico to the USA; Turkey to Germany). The data sets are for origin areas which have all accounted for a large proportion of permanent settlement in the destination areas over the past forty years. In all three immigration countries the proportion of trade from the immigration source grew over the time under consideration, but the absolute and relative scale of immigration varied considerably. The conclusion to the empirical analysis is that no clear correlation exists between migration and trade flows (OECD 1992: 42). This is scarcely a startling conclusion given that the legal controls on immigration in all three cases were dominantly influenced by social considerations relating to constructions of citizenship in the destination countries. By contrast the volume of imports has been affected by the exporting countries ability to respond to international demand for particular products, internationally agreed trade agreements and international competition between producers.

Many of these factors do not influence the relation between trade and international migration within major economic blocks (such as the EU and NAFTA) in the same way as they do interactions between member and non–member states. However, cultural and invisible economic barriers operate to ensure that skilled international migration cannot be "read off" either from indicators such as interstate trade within a trading block or from the extent of internationalization of business activities in an area. For example, a study of attitudes towards the strengthening of the European Community, carried out amongst 1539 professional and executive staff selected from companies in Britain, France, Germany, Italy and Spain, showed distinct national variations in attitudes to the prospect of working in another EU state, even though more than 75 per cent of respondents looked favourably on moves towards the strengthening of the European Community and less than a quarter thought it would lead to negative consequences for "their" economy (UCC 1991). Furthermore there appeared to be no relationship between how well placed staff felt their company was for the enlarged market opportunities and the probability of themselves working in another country. Thus, despite the potential increased economic integration between EU states and the absence of political barriers to staff transfers between countries, less than 16 per cent of Germans and only 30 per cent of Spanish respondents could imagine themselves working in another EU state. By contrast cultural and other constraints on international mobility appeared to be relatively low in Italy where 65 per cent of respondents thought it probable that they would work in another EU state.

A survey of French investments in the UK by Boyle et al. (1994) provides an empirical lens that further illustrates the lack of any direct correlation between increased international economic linkages and increased immigration. The survey covered some 142 French firms that had made investments at no more than three locations in the UK. Table 8.2 shows that only 32 per cent of these companies had chosen to accompany their investment by transferring a high level member of staff from France to the site of their British investment. This finding raises several simple yet important and interrelated questions, such as why international capital flows are not associated directly with human capital flows, and under what circumstances do international investors choose to use expatriates?

Table 8.2   French firms transferring managerial staff to the UK.

		Number of French managers transferred						
Investment type	N	0	1	2	3	4	5	5+
Sales and market distribution	71	52	11	5	3	0	0	0
Producer and related services	71	45	12	3	3	6	1	1

Source: Boyle et al. (1994).

Part of the explanation of why many firms in the survey did not transfer French managers to monitor their investments in the UK can be found in the nature of the investments. Many of the French investments in the UK covered by the survey were buy-outs or take-overs of existing British companies. A similar pattern would be found for other international investment flows between EU countries, with only a minority involving foreign capital setting up new industrial or service units. The need for expatriate managers is clearly less where an existing management structure exists. As Table 8.2 also shows, half of the French investments covered by the survey were in service and market orientated activities and, in line with McPherson's work reported above, it is not surprising to find that this category of investment made use of fewer expatriates. Even where expatriates were used it was less likely than in producer-related companies that more than one French manager would be needed.

Despite these partial explanations of Table 8.2, it remains evident that even where a production-related investment was taking place few companies used expatriates, preferring to use locally recruited staff or personnel already in post at the site of a take-over. Readers interested in further details of the findings of this survey should consult the paper by Boyle et al. (1994), which theorizes the findings in greater detail. But from the perspective of this chapter what is of significance is the empirical support which can be drawn from the study for the view that increased economic integration between EU countries does not automatically lead to proportionate increases in international skilled migration. Consequently, future trends towards increased trade as well as towards more profound and complex integration of EU states cannot be assumed to be a basis in themselves for expecting significant increases in international migration.

Cultural constraints and future skill mobility

Having briefly considered economic arguments in the previous section, the chapter now considers the role of cultural and social factors. These appear to be operating in a contradictory fashion by simultaneously promoting and discouraging skill exchanges between the EU countries. This contradiction has become possible because of the operation at one and the same time of policies facilitating short-term moves and of other policies discouraging permanent settlement and integration. For example, the ERASMUS scheme for student exchange is a case of a policy designed to encourage short-term international mobility of one talented group in the population. It has widespread support on the grounds that it increases international awareness and co-operation. At the same time deep-rooted cultural values have become more resistant, in an era of heightened nationalism, to virtually all visible forms of permanent immigration, and as a consequence across western Europe restrictive immigration policies towards entrants from outside the EU have been reinforced by less visible, but equally potent barriers to labour market entry. The need to belong to national professional bodies or to have foreign qualifications recognized or accredited by local professional associations provide examples of the hidden forms of "ring-fencing" which remain around many skilled labour markets. These less visible barriers have affected intra-EU movements just as much as they have entrants to the EU from other countries.

Table 8.3 illustrates some of the reasons given by professional and managerial staff for resisting attempts by their employer to shift them to a post in another EU state (UCC 1991). It is clear that linguistic differences remain a strong barrier to movement, even amongst highly qualified staff. By 1995 there were eleven official languages in the EU with several others vying for recognition. Each language links to a different culture, producing in Europe a multicultural situation which limits international migration not only through the constraints imposed on communication through the language barrier, but more profoundly through the closure to external penetration which language provides for certain culturally bound social and economic systems.

Also important are certain life-cycle events such as children's schooling that operate to deflate interregional migration, but which become even more potent when international differences in the educational system are introduced to the calculus of the potential migrant. Deroure (1992) found this to be the single most cited constraint in his survey of expatriate usage by large European companies. He linked this to language factors rather than to differences in the curriculum between one country and another. "Parents still seem to attach much importance to their children being educated in their own language" (Deroure 1992: 42).

Variations between countries in the perceptions by potential migrants of the effects of international displacement are also interesting. Table 8.3 indicates, for

Table 8.3 Reasons perceived as inhibiting emigration to other EU states (% of respondents).

	Poor knowledge of foreign language	Children's education	Pension problems	Spouse's career	Lack of recognition of qualification
Germany	18	5	3	5	21
UK	57	45	32	28	23
Spain	55	25	9	19	24
France	33	23	15	32	21
Italy	42	22	18	14	34

Note: Multiple responses: thus percentages do not sum to 100.
Source: Adapted from UCC (1991).

example, that for Italians a lack of recognition of their qualifications was perceived to be a far greater problem than for any other group. The French along with the British were concerned about the effect of international moves on the career of spouses. Too much weight should not be placed on any one statistic in migration perception studies of this kind, given the methodological problems of data capture and the philosophical difficulties of knowing how to interpret such data. What is significant, however, is that Table 8.3 makes it evident that a wide range of cultural and social constraints continue to limit international migration despite the claim of a "Europe without frontiers".

It has been suggested that the high proportion of managerial and professional staff in the EU who would reject an invitation to work elsewhere in the community for their company is one reason why so few companies have vigorously pursued this form of development. Instead they have opted for recruiting local staff to perform management tasks within their own culture zone. In this respect European companies appear to be different from both American and Japanese transnational corporations. Although in global terms the 1980s has seen a shift towards an increased internationalization of the managerial boards of large corporations, within Europe large groups have bucked the trend, keeping their management boards very "national" (Deroure 1992: 26).

In summary there appears to be strong cultural resistance to skilled international migration in northwestern Europe. Although this is less severe in Britain and the Netherlands than in France, Germany or Belgium, it remains true that the opening of a single market in 1993 has had minimal effect in reducing the influence of cultural boundaries on the levels of skilled staff movements between EU states.

The future of highly skilled migration between the EU states

Available evidence suggests that the economic integration of EU states will continue to progress over the decade at a considerable pace, but that this will not be at the expense of any major erosion of cultural differences between member states. Indeed, it seems highly likely that these differences may by the year 2000 have become more important – being celebrated in some quarters and contested in others. Economic integration will mean that marketing and sales interpenetration of national economies by representatives of companies from other member states of the EU will be maintained. This will continue to be done initially by business trips and later, once companies have established local contacts, through the use of marketing staff indigenous to the market concerned. Sales interpenetration will not therefore be a major stimulus to expatriation.

Internationalization of companies within western Europe seems likely to continue to operate through take-overs and buy-outs of local firms, rather than through the creation of many totally new production facilities. The reason for this is that firms wishing to expand production within the EU are not doing so because of wage differentials. If this was their primary motive they would be much more likely to persist with the process of global shift that has led so many companies to relocate production units to developing countries. The formation of joint ventures within Europe has been much more concerned with achieving strategic alliances and economies of scale in rapidly globalizing industrial sectors, while the wave of take-overs has been linked to a desire to increase market share and to position production units advantageously relative to key national markets. In both cases some expatriate presence may be needed to supervise the integration of geographically disparate branches within one organization, but with the passage of time this necessary technical function may be replaced by expatriation for formative reasons, as companies move to train their staff in a wider international context. This was already evident in the survey of French companies operating in the UK (Boyle et al. 1994) where a substantial number of French expatriates covered by the survey proved to be young trainees, who had volunteered for a foreign posting with a French firm in place of carrying out national service. Similar alternative national service possibilities also exist for Italians. A consideration of the likely evolution of economic forces does not therefore lead to a prediction of a major surge in skilled international migration within companies, but to a steady and slow rise. What will take place is a rapid increase in the number of EU citizens working in their own country for companies from other parts of the EU. This will occur not only at the level of semi-skilled but also at the managerial and professional level.

Political strategies favouring short-term international skill mobility also seem likely to prosper once the recession of the early 1990s passes. Schemes such as France's *"coopération civil à l'étranger"* and the EC ERASMUS programme have resulted in significant international population flows in the 1980s, and although expensive to introduce, these schemes have been popular, both with those who have had the opportunity to participate in them and also in host countries. Significantly these schemes, and others like them, have been presented to the public as promoting "mobility" rather than "migration". Not only are such exchanges seen as temporary, but they involve young persons (usually without dependants) who pose no threat to West European societies' defence of the boundaries of national citizenship. They are also not constrained by the family-related factors which have limited the mobility of older cohorts (since they have fewer problems associated with the dual career household, children's education and so on). It is entirely possible that this category of skill exchange will continue to expand at a time when cultural controls seem to be moving towards a stronger limitation on more permanent forms of population redistribution.

It would appear that as the EU states move towards the new millennium, national variations in international migration propensities will remain marked. The positive step of reducing the economic barriers to skill transfers within the EU will not in itself produce an increase in professional and managerial migration. This awaits a time when cultural boundaries become more permeable and when the cultural construction of the meaning of migration shifts to more open ground.

Acknowledgements

The author is very grateful to Mark Boyle, Eva Lelièvre and Ronan Paddison for permission to reproduce the material in Table 8.2 and for their major input in advancing his understanding of French managerial migration.

CHAPTER 9

East–west movement: old barriers, new barriers?

PAUL WHITE & DEBORAH SPORTON

Introduction

The series of political revolutions in eastern Europe that started in the summer of 1989 and ran through to the break-up of the Soviet Union two years later took many commentators by surprise. Few had envisaged political change in as massive and as rapid a fashion as actually transpired, and the period since has seen an outpouring of political and economic writing aimed at repairing the omissions in the predictive gaze through retrospective analysis, and at uncovering the main determinants of the new world.

It should be remembered that one of the catalysts for political change in eastern Europe was Hungary's decision, in the summer of 1989, to open her borders with Austria, thereby allowing the migration of thousands of East Germans who had been camped in Hungary awaiting just such an eventuality (White 1994). There is an irony in the fact that this decision, which led almost inexorably to the opening of the Berlin Wall on 9 November of that year, should affect population movement. Throughout the years of the Cold War the lack of freedom of movement of the population was one of the features of eastern Europe most criticized by the West. The argument of this chapter, however, is that the West is increasingly unprepared to accept the consequences of the granting of such freedom of movement in eastern Europe, and that what we are witnessing in the mid-1990s is the reimposition of border controls, but this time with the enforcing agents being western European rather than eastern European institutions. Nevertheless, there is likely to be some considerable movement between eastern and western Europe into the next century, but the nature of such flows will be specific and not random. This chapter considers the circumstances that will control such movement, the channels through which flows might occur, and the possible outcomes of both actual and latent demands for movement.

The creation of an issue

During the period 1989–91 the theme of potential migration from eastern Europe rose to considerable prominence in public consciousness throughout western Europe, with extensive media discussions in countries such as the UK, France, Belgium, Italy and, inevitably, Germany. Two particularly visual aspects of migration can now be seen as "significant events" in the establishment of such migration as a potential "problem".

The first is the series of television images of intended German migrants encamped near the Hungarian-Austrian frontier in the summer of 1989, demonstrating what at the time was a latent demand for access to the West. The second, and arguably more significant, is reports related to the attempted mass movement of migrants from Albania to Italy during the summer of 1991, when extensive press coverage highlighted the numbers involved, their state of absolute poverty, the mass of arriving Albanians thronging the ports of Bari and Brindisi, and the desperate measures (including commandeering ships) that the migrants were prepared to undertake in fulfilment of their goal. A picture of the migrants swarming over a cargo ship was later used as an international advertisement by the Italian clothing manufacturer Benetton, ensuring that the image became an everyday one throughout Europe. Ironically, a greater number of Albanians had arrived in Italy earlier that same year via controlled movement, and many of those arriving in August 1991 were repatriated anyway (dell'Agnese 1993).

These events creating a public perception of the potential of mass movement were paralleled by political discussion identifying the scale that such movement could take. Organizations in both the west and the east came forward with estimates of the numbers of ex-Soviet citizens who might seek to move to the west over the next decade, with results ranging from 2 to 20 millions. Opinion polls were conducted in various eastern European states to identify possible movers and destinations. Table 9.1 shows the results of one such survey conducted in Hungary, Czechoslovakia (as it then was), Poland and the Soviet Union (in fact the entire sample was of Muscovites) in the spring of 1991. Such surveys have turned out to be fanciful as predictors. In the three years since the data in Table 9.1 were collected there is an implication here that, taking only those with definite plans to move abroad, there should have been over 1.5 million departures of Russians, over 600000 from the Czech Republic and Slovakia, over 700000 from Poland, and 100000 from Hungary. Such flows have simply not occurred.

However, the survey data shown in Table 9.1 are indicative of attitudes even if they are not usefully predictive. In particular, the potential destinations are of interest. Three of the top four places are taken by traditional global migration destination areas in North America and Oceania: amongst this most favoured destination status only Germany features as a European case, and here the attraction

Table 9.1 Expressed migration desires amongst Eastern Europeans, 1991 (%).

	Hungarians	Czechoslovaks	Poles	Russians
In the next three years, do you intend to go and live abroad?				
Yes, definitely	1	4	2	1
Yes, perhaps	3	13	4	9
No, probably not	7	23	20	19
No, definitely not	88	59	69	67
Don't know	1	1	5	4
If you had to live abroad where would you go? (more than one answer allowed)				
USA	27	31	47	38
Canada	25	45	37	21
Germany	34	40	25	17
Australia/New Zealand	26	25	20	21
Austria	28	37	9	9
Switzerland	24	26	11	21
France	16	16	18	23
Scandinavia	15	14	14	22
Italy	14	6	14	15

Source: Guardian, 24 May 1991.

for Russians is much less than for those from other parts of Central Europe. Austria is of attraction for citizens of the countries that were once part of the Austro–Hungarian Empire – suggesting the significance of historical links, and underlining the fact that one of the more significant effects of the velvet revolutions in central Europe was to create the conditions for Vienna to re-emerge as the capital of Central Europe as it had been before the First World War. Neutral Switzerland was seen as a desirable destination, but amongst the other western European states only France, Scandinavia and Italy had any support. In each sample less than 5 per cent named an eastern European country as a favoured potential destination.

Two further comments can be made on the basis of this information. First, many of these expressed preferences were clearly made with little real understanding of actual opportunities for migration (and no such understanding was sought by the survey). In particular the actual costs of movement seemed to play no role in preferences, and those costs have become prohibitive through the rapid fall in the purchasing power of most eastern European currencies over the past few years. Secondly, the western European states with most to "fear" from such surveys both in terms of their perceived desirability and their proximity and ease of access for migrants are Germany, Austria and, within Scandinavia, Finland.

Political responses to such surveys have been remarkably similar in tone across Europe – to deplore the possible scale of movement. In the West there has been the linking of potential migration, through refugee flows, to local racism in a nexus of opinion generation concerning the need for stricter controls: in the East there have been concerns expressed about the draining of skills and talent from poor economies as they face up to market forces. There has also, however, been the suspicion that some of the more alarmist estimates of numbers of migrants have been fuelled by eastern European politicians seeking Western aid as a counter to such moves and looking for the sort of Western response encapsulated in an editorial in *The Independent* newspaper in July 1991 under the headline "The Soviets Must Be Helped". The wider moves towards European Union immigration policies will be mentioned later in this chapter.

Movements before 1989

One of the problems in coming to terms with population movements from East to West in Europe in the new order of the 1990s onwards is that there appears at first sight to be little antecedent movement on which new patterns can be built, and little in the way of established findings on pre-existing migrations within eastern Europe itself. Even were such findings to exist it could be argued that they would be poor indicators of migration propensity in situations involving new political and economic structures, especially as regards employment. In fact, antecedent movements have been of significance, and the past can yield certain clues to the future, even in this field.

Certainly there were strong controls on movement within the command economies of eastern Europe, but urbanization movements have been a continuous feature of the post-war period within countries. Within the former Soviet Union such movements often had a strong ethnic dimension to them. This resulted partly from the ethnic diversity of the country, and gave rise to ethnic minority distributions within certain of the major cities, as for example in the former Leningrad where a geography of ethnic residence was apparent, just as in cities in the Western world (Starovojtova 1987). Throughout many of the Union republics, however, one of the most significant elements of migration involved the movement of Russians, as a result of which the previous relative ethnic homogeneity of certain republics, particularly in the Baltics, was weakened by the rapid growth of a predominantly Russian residential element (Karger 1988, Rowland 1988). Between 1979 and 1989 the rate of growth of the Russian population in the Ukraine, Belarus, Lithuania, Latvia and Estonia exceeded its growth within the Russian federation itself. Only the Belarussians and the Ukrainians were showing any significant reverse flow into the Russian Federation (Zayonchkovskaya et al. 1991). By 1989 amongst all the then Soviet republics only the Russian Federation, Azerbaijan and Armenia had over 80 per cent of their populations drawn from the indigenous nationality. Less than two-thirds of the population was ethnically native in several republics – Latvia (53%), Estonia (62%), Kirghizia (52%), Tadjikistan (62%) and Kazakhstan (40%). In all the European republics except those of the Caucasus the period 1979–89 had brought greater ethnic mixing, in all the other republics the trend had been the other way, towards greater homogeneity. As we shall see, these ethnic dimensions of Soviet migration are of some significance for the future.

Elsewhere in eastern Europe internal migration has also occurred throughout recent decades, particularly affected by regional economic growth and labour requirements (Compton 1976, Korcelli 1988), although ethnic issues have also been of importance in many areas (Satmarescu 1975).

However, international migration was at a very small scale in eastern Europe in comparison with western Europe. Certainly COMECON did organize some labour interchanges (Guha 1978) but these were on a very small scale. Instead there was some organization of labour migration from Socialist Third World countries such as Cuba, Mozambique and Vietnam into the more industrial economies of Czechoslovakia and East Germany. These migrants, brought in for comradely solidarity, were either quickly repatriated during 1989–90 or were left to their own fates in increasingly racist environments. The former Yugoslavia, which was effectively part of the western European migration field, is omitted from this discussion.

We should note, however, that during the 1980s there were certain migration streams out of eastern Europe, each of them operating under considerable governmental control. The most important of these can be summarized as:

- A controlled outflow of older people from East to West Germany. In 1987 and 1988, the last two complete years before the fall of the Berlin Wall, West Germany had gained a net 61 206 migrants from its eastern neighbour.
- The outflow of those of German ethnicity (the Aussiedler) from several eastern European countries to the Federal Republic. The largest of these flows was from Poland (800 000 of the total of 1.2 millions between 1970 and 1989), the Soviet Union and Romania (Guillon 1989, Salt 1993). The scale of these flows, and the somewhat tenuous claims on "Germanness" for some of the migrants, has become a major discussion point in the reunited Germany, and has played a role in the current debate on German citizenship (Herdegen 1989, Jones & Wild 1992, Treibel 1994).
- A highly fluctuating but controlled outflow of Soviet Jews (Heitman 1987). Between 1987 and 1989 there were 98 000 Jewish emigrants from the USSR. However, during 1990–91, preceding changes in laws on exit permits and conditions, there were 340 000 arrivals in Israel from the Soviet Union (Rowley 1992), with much smaller numbers going to the USA and Germany. Within Israel they have become particularly associated with Jewish settlement on the West Bank.

- A smaller continuing flow of Greeks from Black Sea settlements to Greece, amounting to approximately 15 000 per annum (Chesnais 1991b).

These are recent flows. However, it needs to be born in mind that historical movements have created considerable communities of eastern European origin in many other parts of the world. Poles and Armenians are particularly widely distributed, but there are also significant emigré groups of most other nationalities in many parts of western Europe, in North America, in Australasia and even in Latin America (particularly in Argentina). Many of these communities trace their history back to migration at the turn of the twentieth century, but others were created by population displacements relating to the two world wars.

In total these international flows do not amount to a great deal. They have also taken place under particular conditions in which ethnic reasons underlain by political considerations have been both the driving force behind movement and the controlling influences on it. Demographers and population geographers, although lacking in many earlier migration flows on which to base their predictions, are nevertheless able to apply many of their existing concepts to discussion of possible new migration flows in the 1990s and beyond. Amongst these concepts can be counted views of migration as human capital mobility, operating within the parameters of neoclassical views of regional economic disparities; ideas on the responsiveness of potential migrants to various types of information flows; concepts concerning the existence and efficacy of migration channels which facilitate movement; notions of chain migration and of return migration; and established theories on the effects of migration policy. The contexts for international population mobility from eastern Europe may be new, but many of these older analytical tools can be pressed into service.

The context for future movement

The circumstances of the various countries of eastern Europe are relatively diverse, despite their general emergence from command economic to capitalist structures. Discussion here will therefore be primarily concerned with the former Soviet Union, with occasional reference to the situation elsewhere.

A first major feature of the ex-Soviet Union is the expectation that over the course of the next twenty years we shall see the slow integration of the economies of the new republics into the European economy as a whole, as part of the European semi-periphery (Gritsai & Treivish 1990). This integration should be expected to bring about adjustments to regional labour markets and economic structures not just in the former Soviet Union but also elsewhere. In terms of neoclassical economic theory, this new opportunity for the European (or global) integration of these eastern economies should result in several new flows. First there should be flows of capital from the Western economic powers, given the attractions of investment in low wage economies, and the possible further attraction of the continuation of low environmental controls on production activity. These capital flows would primarily affect, in the first instance, manufacturing activity. The output would be targeted towards world markets but, as local economic growth occurred in the east, increasingly also at a newly developing internal market. Modelling the West–East capital flows needed to make significant impacts on the ex-Soviet economies, given their size, suggests that the economic differential between East and West in Europe will scarcely be eroded over the next decade or so (Öberg & Boubnova 1993).

Some of these theoretical suggestions are already developing into reality. We might note that many larger Western manufacturers are now opening production units in the East: to take only the motor manufacturing sector, Fiat are expanding their existing production at Yelabuga and Togliatti to the east of Moscow, as well as at Bielsko Biala in southern Poland; General Motors have been negotiating to build a new plant at Warsaw and have increased production at Szentgotthard in Hungary; Volkswagen have made further investments in Mlada-Boleslav in the Czech republic and at Bratislava in Slovakia; Suzuki have opened a plant in Hungary (Sadler et al. 1993). The Czech economy actually recorded a trade surplus for 1993 (*The Guardian*, 24 February 1994), and the privatization of the economy there, in Poland and in Hungary has been reflected in massive increases in stock-market values of companies. However, Western investment requires confidence in political stability, and it is that confidence that is lacking in certain other eastern European states, and particularly in Russia and certain other former Soviet republics.

However, West–East capital flows into an integrated semi-peripheral economy should be paralleled by equilibrating East–West labour flows from low wage to high wage economies. It is on this point that economic analysis shows its limits, since such labour movement, occurring officially or on any large scale, has become anathema to most political parties in western Europe.

A second contextual feature of contemporary eastern Europe concerns income disparities. In the full employment (involving significant underemployment) of the old command economies income differentials were low, certainly in comparison with the rest of Europe. Privatization, the market economy, competition, and substantial inflation in some countries all conspire to increase unemployment, to increase the income of certain entrepreneurs and groups, and to increase the gap in living standards between the most prosperous and the worst off. These could be the conditions leading to a mass movement of effectively economic refugees seeking any sort of livelihood elsewhere, particularly in a scenario of the breakdown of the economy and of distribution networks. Such a scenario has been planned against by Finland, which has feared the arrival on its doorstep of a million or more starving

Russians (Finland's Karelian frontier with Russia lies less than 200 kilometres from St Petersburg). Finland has created a network of skeletal emergency camps along this frontier and has stockpiled food that could be used to feed migrants on the frontier without giving them access to Finnish territory. This may be seen as a "worst-case" scenario affecting a natural "buffer state" against westward migration from the former Soviet Union, but given the economic chaos that affects Russia today such planning is not overcautious.

A third contextual factor of very great importance throughout eastern Europe is the potential for ethnic disturbance and consequent migration away from violence or persecution. The example of the former Yugoslavia is a salutary reminder of the way in which ethnicity, largely kept submerged as an issue under the Communist regimes, has re-emerged as a critical destabilizing element in regional politics. Latest figures from the United Nations High Commission for Refugees (May 1994) suggest that 4.562 million people have been displaced by unrest in ex-Yugoslavia. However, it is notable that, as a result of a reluctance on the part of other European governments to accept asylum-seekers from the region, 84 per cent of these displaced persons have remained within the former boundaries of Yugoslavia: of the 750000 who have gone elsewhere, the largest contingents have gone to Germany (403000), Sweden (59000), and Austria (55000): the figure for Hungary (28000) is almost certainly an underestimate, since many ethnic Hungarians from the formerly autonomous province of Vojvodina have simply moved over the Hungarian border without any official notification (Morokvasic 1992, Redei 1992).

Much of the rest of eastern Europe has the potential to generate other ethnic refugees, although not necessarily on such a scale. The fragmentation of the USSR, in particular, has already given rise to many conflicts (Glezer & Straletsky 1991), among them the war between Armenia and Azerbaijan over Nagorno–Karabakh, civil war in Georgia over the attempted secession of Abkhasia, trouble in North and South Ossetia, and unrest in Moldova where the whole question of "national" identity and relationships with Romania and the rest of the former Soviet Union has been particularly volatile (Miller 1994). Figure 9.1 provides an illustration of potential ethnic trouble spots in the western part of the Commonwealth of Independent States, ignoring the Asian republics where there is a considerable possibility of unrest (Smith 1990, Naumkin 1994).

Many of the problems relate to longstanding historical developments, but others reflect the pattern of recent Soviet migration outlined earlier in this chapter. There are many moves by newly independent republics, particularly the Baltics, to overthrow what is seen as invasive Russian hegemony by restricting the rights of immigrant Russians (for example, through refusal to grant citizenship or by an insistence on the official use of the "local" state language), in the hope that this will induce Russians to "go home". Lithuania has launched a campaign to induce the Russians to give up their control of the enclave of Kaliningrad, and elsewhere the rhetoric

149

Figure 9.1 Areas of potential ethnic conflict in the former Soviet Union. *Source:* redrawn from O. Glezer et al., "A map of unrest in the USSR", *Moscow News* **11**, 1991.

of identity and nationalism has been sharpened up in many ways, with people from other former Soviet republics now being labelled as "foreign-born" (Sakkeus 1992).

Elsewhere throughout eastern Europe there are similar tensions: over the Hungarian minorities in both Slovakia and Romania; over the Polish-Lithuanian border region; over Turks in Bulgaria; over Greeks in Albania; and in parts of Yugoslavia not so far affected by fighting, most notably Macedonia. Commentators have noted the rise of an extreme-right populist nationalism in several former Communist states, exemplified by Zhirinovsky in Russia but also becoming a significant political force elsewhere (Hockenos 1994, Laqueur 1994). Actual and potential ethnic conflicts add a potent ingredient to any consideration of future migration and refugee pressures in eastern Europe. Conflict in, for example, the Baltic States may lead to a Russian refugee problem within Russia; but conflict in certain other regions (such as Moldova or in the Caucasus) could lead to international refugee flows of considerable size and complexity.

A fourth contextual factor relates to the possible migration channels along which any new large-scale movement could occur. The concept of migration channels highlights the interconnections between information flows, organizational structures and migration mechanisms that may act together to facilitate the mobility of specific population subgroups (Findlay 1990). Much of the established migration literature adopting this concept relates to high level migration, often within the internal labour markets of international firms, but the concept can be applied more widely.

In the context of migration from eastern Europe, given the effectively closed frontiers of the recent past, few effective migration channels exist. With the exception of the governmentally controlled flows mentioned above, there have been very few primary migrants to facilitate later moves through chain migration systems. No internal labour markets have linked eastern Europe with the rest of the world within multinational private enterprises.

The only possible existing migration channels, in the normal sense, affect intellectuals, artists, academicians and others of elite groups who have in the past been permitted to travel in the West and who have been part of international networks. Major figures from these worlds have in the past defected to the West or been expelled there, as in the cases of Rudolph Nureyev and Alexander Solzhenitsyn, but many others have had contacts with the West. Over the past few years there can be very few Western university teachers with contacts in eastern Europe who have not been asked by eastern European colleagues for advice about getting a job in the West. The numbers who will actually succeed in moving will probably be very small, however, and it is notable that various Western aid projects (such as the COPERNICUS scheme for academics, and programmes for retraining ex-Soviet defence workers) have been aimed at stemming skilled migration flows which are conceptualized as haemorrhaging crucial workforce elements from central and eastern Europe.

However, on a much broader definition of potential migration channels, there have been signs that various eastern European emigré communities elsewhere in the world would be prepared to support would-be migrants from their original homelands, especially in any future scenario of economic crisis or ethnic conflict.

Table 9.2   Citizens of Eastern European countries resident elsewhere, 1993.

Resident in	From (figures in thousands)					
	Bulgaria	Czechoslovakia	Hungary	Poland	Romania	Ex-USSR
Austria (1991)	3.6	11.3	10.6	18.3	18.5	2.1
Belgium	nd	0.6	0.7	4.8	nd	1.2
Denmark	0.2	0.4	0.3	5.0	0.1	0.2
Finland	0.3	0.2	0.4	0.7	0.3	12.1
France (1990)	0.8	2.0	2.9	46.2	5.7	4.3
West Germany (1992)	32.7	46.7	56.4	271.2	92.1	52.8
Italy (1992)	1.3	1.2	1.5	9.1	5.2	2.3
Netherlands	nd	nd	1.2	5.3	1.8	2.1
Norway	0.4	0.2	0.2	2.9	0.3	0.9
Spain	0.6	0.4	0.2	3.2	0.7	1.2
Sweden	2.1	1.7	3.5	16.4	5.1	4.0
Switzerland	0.8	5.7	4.5	5.4	2.7	2.5
UK (1989–91)	nd	1.0	3.0	34.0	nd	13.0

nd = no data
Source: Council of Europe 1994, *Recent demographic developments in Europe, 1993.*

In this respect the size and distribution of such communities elsewhere is of some interest. Data from the Council of Europe on stocks of foreigners in 1993 (Table 9.2) actually conflate two very different types of migrants. First there are the long-standing groups of residents, many of whom have retained a "foreign" nationality for several decades. Prime amongst these are the Poles, who have origins in western Europe going back in many cases to labour movement before the Second World War (Treibel 1994), or in other cases, such as the United Kingdom, became established as an emigré community in the aftermath of Communist take-over in eastern Europe after the war. Certain other populations in Table 9.2 originate in political flight during the post-war period: for example, the numbers of Czechs and Hungarians in Switzerland have actually been diminishing steadily since the initial influxes in 1968 and 1956 respectively.

Secondly, elsewhere the numbers of eastern European citizens are increasing rapidly as a result of recent arrivals. These figures are again likely to be underestimates since they come from official data and ignore the presence of clandestines in the migrant population. In terms of destination countries, the pre-eminent place is taken by Germany (the data here refer only to West Germany, since data collection in the former German Democratic Republic was much less accurate), followed by Austria. Finland was also significant for citizens of the former Soviet Union (echoing a point made earlier about the vulnerability of both Austria and Finland to eastern European in-movement), with Sweden of some significance for recent Polish movement, linked by ferries across the Baltic Sea.

Overall, however, with the exception of the Poles, these figures are relatively small: the numbers of usable migration channels linking these emigré populations and their "homelands" is likely to be relatively small. However, these data relate to populations still retaining their eastern European nationalities in a situation where many older emigrés have naturalized to a new nationality. If instead we consider the numbers of people in eastern Europe who are ethnically linked to groups elsewhere, then the potential for facilitated migration can be seen to be considerably greater. This ethnic linking could be of significance if ethnic unrest increases.

We shall concentrate the discussion here primarily on the former Soviet Union. Öberg and Boubnova (1993), using censuses from various countries around the world, in some of which questions relate to ethnic origin rather than simply nationality, have calculated that there are perhaps 1.7 million "Russians" (including Belorussians) elsewhere in the world, 1.3 million Ukrainians, 1.3 million Armenians, 0.7 million Azerbaijanis, as well as smaller numbers of Lithuanians (350000), Georgians (120000), Estonians (80000) and Latvians (70000). Chesnais (1991b) has used the last Union census (1989) to estimate the populations of the former Soviet Union who have ethnic connections elsewhere (see Table 9.3). The groups with links to the rest of Europe (ignoring internal ethnic distributions within the former Soviet Union) total 10.75 millions. Certain groups would be unlikely to play any real role in population change elsewhere in Europe, most notably the Jews who, as noted earlier, were migrating to Israel in large numbers between 1989 and 1991 during which period they enjoyed privileged rights of exit from the Soviet Union. However, even here it is possible that internal developments within Israel (for example, through greater autonomy in the West Bank) could divert existing Jewish settlers and future potential migrants back into Europe.

Some comment is useful on the other groups. The Armenians represent a long-standing diaspora, largely related to the massacres and upheaval in the region during the First World War (Ter Minassian 1989). As a result there are significant Armenian communities in several western European cities, for example Manchester (Mason 1978) and Paris (Hovanessian 1988). These, however, are very long-standing communities with little direct contemporary contact with Armenia, although with very strong emotional ties.

The two million German residents have created considerable discussion since the establishment of the Commonwealth of Independent States, involving President Yeltsin in negotiations with the German government, with other countries (for example Argentina) coming forward as possible resettlement locations. Much has been written about resettlement costs, either within the Russian Federation or elsewhere (*Der Spiegel*, 24 June 1991), and although the German Basic Law (Grundgesetz) clause 16 guarantees their right of entry to Germany it is clear that

Table 9.3 Residents of the former Soviet
Union (1989) with ethnic links elsewhere in
Europe (in thousands).

Armenians	4623
Germans	2038
Jews	1449
Poles	1126
Bulgarians	374
Greeks	358
Karelians and Finns	178
Hungarians	171
Romanians	146
Slovaks	16
Czechs	9
(Gypsies	262)

Source: Chesnais (1991b: 10).

the German people and government do not want them, at least, not if they arrived
by mass movement. In 1992 Yeltsin moved to re-establish an autonomous German
republic in the Volga region, defusing the situation for the time being.

The figures given in Table 9.3 for those with Polish, Karelian, Finnish, Hungarian, Romanian and Slovakian connections outside the CIS largely relate to the
drawing and redrawing of political boundaries through mixed ethnic areas during
the twentieth century, although some elements of labour migration have also been
involved. The inclusion in Table 9.3 of a figure for gypsies is a double reminder that
these have been one of the persecuted minorities in the former Soviet Union (as
elsewhere) and that the largest concentration of gypsies in Europe is in Bulgaria,
with a substantial population also in Romania: however, it is doubtful whether
these countries would welcome any further gypsy arrivals.

These figures illustrate the point that potential migration connections exist, but
at an extremely tenuous level. Migration paths and channels are not properly established, and this in part explains the failure of large-scale movement to take off since
1989, despite the intentions expressed earlier in Table 9.1. However, it is the fifth
and final contextual point that also explains the fact that threats of mass movement
have not translated into actuality.

The opening up of the East came at a time when migration was already on the
political agenda in western Europe, but in a rather different context. The Schengen
agreement had been initiated in 1985, at a time when, as Salt (1993: 41) has
observed, the European Community "could look forward to a generally imper-
meable barrier to emigration from the Warsaw Pact countries". Schengen and its

extension as more countries signed up, was a vision of open frontiers within the European Union, with the prospect of the emergence of a common immigration policy relating to other countries. As the perceived threat of large-scale movement from outside western Europe (including movement from both the East and the South; see Ch. 10) has grown, so individual governments have moved to tighten up their own policies, with particularly public debates over such changes in Germany, France and the United Kingdom. At the same time legislation on asylum applications has been strengthened, with procedures being speeded up in an effort to keep the momentum up such that rejected claimants can be sent back quickly rather than getting a toehold in their applicant countries and being granted exceptional leave to remain. Immigration policies have proved very difficult to "harmonize" since to do so raises crucial questions of national sovereignty in an arena where significant elements of public opinion are becoming more inflamed and racist in their rhetoric (Brochmann 1993, Salt 1993). This particularly applies to policies on asylum seekers and measures taken against those locally identified as "illegal" immigrants. Nor is there any real harmonization of data collection on migration, and this creates a particular problem for movements involving eastern Europe where there are generally no outflow data available to compare in any way with inflow data in destination countries (Willekens 1994).

The net effect of nearly a decade of intense political and public debate in western Europe is an imperfect implementation of the Schengen proposals on complete freedom of movement within the European Economic Area (EEA, made up of the European Union plus the countries of the European Free Trade Association), coupled with more restrictive control on the movement of third country nationals. Certain countries, such as the United Kingdom, have never embraced the Schengen principles, whereas a locationally important player in the European migration scene, Switzerland, voted in late 1992 to stay out of the EEA completely.

At the same time that these developments have been taking place in western Europe, the countries of the East have moved to the evolution of more standard migration policies allowing their citizens the freedom of travel. Most significant among these was the legislation passed in the Soviet Union in May 1991 which was intended to come into effect on 1 January 1993 (Salt 1993). This was overcome by events in terms of the dismantling of the Union and the creation of the Commonwealth of Independent States and the independent Baltic republics. The CIS is intended to be a loose federation without significant inter-republic borders, but there have been signs that controls could be implemented: in October 1993, for example, the mayor of Moscow moved to require residence applications from citizens of CIS republics other than the Russian Federation, thereby effectively reimposing a degree of migration control that had earlier disappeared.

Elsewhere in eastern Europe foreign travel for citizens was quickly opened up during or immediately after 1989 (Poland actually legislated a year earlier, in 1988), but they have also, like western Europe, acted more recently to tighten up frontier controls, particularly on their eastern frontiers (Korcelli 1992). At first it appeared that Austria and the Oder River frontier between Germany and Poland were becoming the new front-line locations in a struggle to keep out westward-flowing migrants, but the rapid growth of intended movers from farther east and southeast in Poland and Hungary in particular has lead these countries to strengthen their eastern border controls in turn. This has been supplemented by a series of agreements between the Schengen countries (lead by Germany) and individual eastern European governments imposing the obligation on any government to accept back any migrant who has tried to cross to another state without legal right to do so, with the sending country (invariably in western Europe) adding some financial help to the recipient country. These agreements involve, on the eastern side, Hungary, Poland, Romania and Bulgaria: apart from Schengen signatories they have also been negotiated by Austria and Switzerland. These have dual effects: they tend to make the countries of east-central Europe into buffer states against migration from elsewhere, and they also encourage these states to tighten up their own frontier controls against this "transit migration". Between January and October 1993 the Czech Republic detained 36000 people on its borders; in the first six months of the same year Poland detained 15000 in the same way (UN/ECE International Migration Bulletin, November 1993). Although most of those detained came from elsewhere in eastern Europe, there were also significant numbers of Chinese and Indians travelling as part of organized migrant smuggling systems.

A Guardian newspaper headline has claimed (16 February 1993): "Poland erects new iron curtain to stem flood from the east." It appears increasingly likely that the old iron curtain, erected by the Communist states to "preserve" the East will be replaced by a new set of barriers to "preserve" the interests of the West. Exactly where the new East–West frontier will lie is still open for negotiation. Arguably Poland, the Czech Republic, Hungary and Slovenia will lie to the west of it: Estonia, Latvia and Lithuania may also do so, as may possibly Bulgaria and Romania. The ultimate position of the rest of eastern Europe, and particularly of the CIS is yet to be determined but looks less likely to be on the "right" side for easy migration.

Possible movers

The contextual issues presented above create a series of conflicting migratory pressures: on the one hand, several forces (economic and political) that might be assumed to increase individuals propensity to move, but on the other hand a series of retarding features such as the absence of established flow channels and, above all,

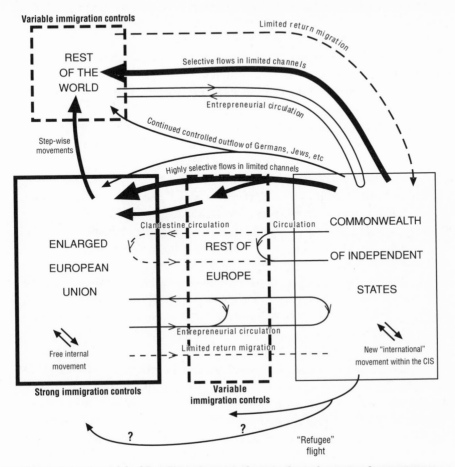

Figure 9.2 A model of East-West migration flows in the early twenty-first century.

the increasing climate of restriction. In these circumstances, what patterns of migration linking eastern and western Europe might emerge over the next twenty years or so? Figure 9.2 presents a schematic attempt to model the answer to this question.

Figure 9.2 takes as its basis certain points already made. These include the fact that the real desires among many eastern Europeans appear to be for settlement in the New World rather than in Europe, so that many moves could be of a step-wise nature. Europe is divided here into three blocks – a progressively enlarging European Union, the Commonwealth of Independent States, and a group of countries labelled here as the "rest of Europe". What these countries might consist of is not defined here: which camp Switzerland will eventually join is not predictable at this stage, nor the speed of absorption of certain eastern European states into the EU.

The further basic element dealt with in Figure 9.2 concerns the strength of migration controls, which are seen as tough and potentially unitary around the EU, weaker and more fragmentary in the rest of Europe. Outside Europe controls will continue to be variable and possibly highly selective.

Given these basic assumptions, and the other contextual points made earlier, the following possible migration flows can be put forward.

- Some continuation of the controlled out-movements of Germans, Jews and Greeks mentioned earlier.
- Highly selected movements of people with particular skills to the West, recruited under strict government control to carry out specified functions in Western economies (Vishnesky & Zayonchkovskaya 1992). Some of this movement may represent a continuation of the earlier movement of "world citizen" artists and intellectuals referred to earlier, but it is possible that with renewed economic growth in the Fifth Kondratieff cycle, and with decreasing population bases in various western European countries, labour import will probably become a more common scenario in the early twenty-first century. In the case of the "rest of Europe", increasing economic strength may progressively bring relaxing of controls towards free exchanges of labour with the EU. For all eastern European countries, but especially those who are accepted into the EU, this will be facilitated by Western investment which will bring workforces in these countries, albeit slowly, into the internal labour markets of Western companies. Some movement may well constitute "brain drain" from the East, although for qualified migrants who accept lower-level jobs this may also involve "brain waste" (Rhode 1991). We should not expect this migration stream to develop overnight: evolution will be slow. By the year 2010 we might expect significant flows of Poles, Czechs, and Hungarians to be occurring within the EU, with possibly other flows involving the "rest of Europe", but much of the absorption of eastern European workers into the internal labour forces of transnational corporations may well still retain that labour in its home area.
- Such Western investment will also create West-to-East flows in what is here labelled as "entrepreneurial circulation", taking Western managers and experts on short-term contracts to work in or set up subsidiary enterprises in eastern Europe. Moscow now has its first private golf club for Western citizens, and other cities and countries will surely see similar extensions of the infrastructure for international executives and families.
- There is also likely to be a small amount of return migration over the next decade or so of people originating in eastern Europe who have lived "abroad" most of their lifetimes but now wish to move back to their birthplaces. Interest in such movement has been reported among, for example, Latvians and Lithuanians in the USA. However, the scale of these moves will almost cer-

tainly be very small: those involved are now elderly, and the disparity in living standards will militate against many giving up relatively more comfortable lives in the West for an uncertain future elsewhere.

These migration streams are all politically "acceptable" ones to the powers of western Europe and elsewhere. What is less "acceptable", and anyway harder to predict, are any more massive and uncontrolled flows. These can be posited to be of two kinds.

- Petty entrepreneurial (often clandestine) movement from East to West. There is much evidence of such movement visible on the streets of many large cities in Europe today. There are an estimated 7 million trips made across Poland's eastern frontier by Russian petty traders each year (*The Guardian*, 16 February 1993). Street trading, busking and begging by Romanians, gypsies, Bulgarians and others is now commonplace in Berlin, Vienna, Prague and Budapest. Withdrawing ex-Soviet military personnel have absconded and stayed on in several countries, sometimes making a living out of selling old military equipment.

    Throughout Europe clandestine migrants find ways of circumventing border controls, and can make a living in the more fragmented economic and labour structures of post-industrial societies. Although politically unacceptable such migrants may not be without their advantages for their destination countries: early in 1993 the German building industry was reported to be illegally employing up to half a million foreigners (not all of them actually illegal immigrants) at wage rates that drastically reduced construction costs. We know from the period before 1989 that officials can often turn a blind eye to clandestines when they have secured a niche in the labour market (Leitner 1990). Individual or group entrepreneurial moves from East to West will depend in part on the efficacy of Western border controls, but they will also reflect the state of economic structures and of consumer demand throughout Europe. In total the numbers involved can be expected to be quite considerable, particularly if similar inflows through southern Europe are taken into account (see Ch. 10).

- The hardest type of migration to predict will involve flight from economic chaos or, more especially, ethnic conflict and civil war. As pointed out earlier, the migratory effects of the Yugoslav crisis have been very largely confined to the region itself. Whether the effects of large scale conflict in, for example, the Caucasus could be similarly contained, and what the consequences might be for the European demographic scene is harder to forecast.

We must finally address the likely establishment of "international migration" where previously it was "internal".

- Current developments suggest that the CIS will not be a stable commonwealth for some time. The Ukraine electorate has elected first a nationalist and then a pro-Russian as President; Belarus has gone for populism; Moldova has seen internal upheaval. Controls on internal movement within the CIS will create the circumstances under which migration is truly seen as "cross-border", and the significance of this will increase with possible return moves of Russians to their "home" republic (Cole & Filatotchev 1992, Dunlop 1993).

Conclusions

The processes of migratory adjustment to new political and economic (and potentially to new ethnic) realities in eastern Europe will be a long one. With the moves to German reunification in 1989–90 the migration effect took place quickly in a short burst that has subsided to a more regular flow. But German reunification, with all its complexities, was a far more simple development than the exposure of millions of citizens of eastern Europe as a whole, for the first time in their lives, to the possibility of freedom of movement. Migration is seated both within the biographical history of individuals and also within the structural features of the economy and society at large. The trigger mechanisms are complex, and where little real knowledge of opportunities is available elsewhere the translation of dreams into mobile realities is slow, if it occurs at all. First estimates of mass migration out of eastern Europe once the velvet revolutions and political upheavals of 1989–91 had taken their course were grossly overstated.

However, such estimates and fears helped to create the climate of political opinion in the West resulting in the tightening of controls and the restriction of movement possibilities in the future. Restricting access had the immediate effect of increasing the numbers claiming asylum as a means of entry, but now that measures have been taken to limit the numbers of those successful in that endeavour western Europe is truly moving to a highly restrictive immigration regime. The irony of the fall of the Iron Curtain is that beforehand the peoples of eastern Europe had no freedom to leave their own countries: now they have that freedom they have nowhere to go. In a decade's time we may look back on the early 1990s as a period of relative openness of the whole of Europe before the controls are reimposed from a different direction. East–West migration in the future will certainly be greater than in the past, but it will be less than the alarmist predictions and will be highly selective according to skill, initiative or desperation.

CHAPTER 10

Future migration into southern Europe

MAURA MISITI, CALOGERO MUSCARÀ,

PABLO PUMARES, VICENTE RODRIGUEZ,

PAUL WHITE

The migration turnaround in southern Europe

During the boom years of post-war economic growth in northwest Europe the countries of the Mediterranean shores played a vital role in the supply of labour to their northern neighbours. Although there were regional variations within individual southern European countries in terms of contributions to emigration flows (King 1976), nevertheless in each country as a whole net migration flows showed population loss for long periods throughout the 1950s and 1960s, and even into the early 1970s.

Southern Europe seemed to be continuing to play its historic role of several centuries as a labour reserve and area of emigration, providing workers for the rest of Europe as well as contributing significantly to the Europeanization of other parts of the globe, particularly in the Americas.

In many ways the most important contribution came from the most populous country, Italy, with its massive population loss to the Americas at the start of the twentieth century, supplemented by net flows of between 100 000 and 350 000 per annum to the rest of Europe during the period from 1947 to 1970 (King 1993a). Spanish out-movement was at a very high level from the beginning of the century until 1930, with movement of around 100 000 migrants per year to Latin America. From the late 1950s to the 1970s there were a further 60 000 migrants per year to destinations within Europe (Sorel 1974, Gregory 1978, Dirección General de Migraciones 1993).

Portugal, Greece and the former Yugoslavia were also important contributors to international movement, entering the system at different periods and with varying destinations. In the case of Portugal, large-scale emigration continued throughout the last years of Salazar's regime, despite attempts to stop it (Leloup 1972), although the loss of Angola and Mozambique in 1975 brought the very sudden re-entry of around 800 000 *retornados* within a year (Dewdney & White 1986). Greek migration – predominantly but not exclusively to Germany – peaked during the 1960s (Lianos 1975). Two other cases should be briefly mentioned, although they will not be discussed further in this chapter. In what was then Yugoslavia the pattern and significance of out-migration accorded closely with what was later to become the map of territorial contest over the break-up of the federation, particularly in Bosnia (Lichtenberger 1984). Finally, although only partly in Europe, we should note that out-movement from Turkey was on a scale sufficient to make Turks the most numerous foreigner group in West Germany (Martin 1991).

By the early 1970s, however, migration flows were already beginning to change, and the years both before and after the mid-1970s economic recession saw the initiation of large-scale return migration of people who had left southern Europe maybe twenty years earlier. The effects of such moves were considerable in several respects, and a considerable amount of academic interest was generated in these flows, partly summed up in the collected work edited by King (1986). Return migration was still, however, within the generally accepted cycle of labour migration that had been seen as operative throughout the post-war period in western Europe (Böhning 1972). Certainly much of the return was not so much brought about by the fulfilment of objectives at destinations, nor directly by unemployment and the disbanding of an "industrial reserve army", but by improving prospects in southern European countries witnessing relatively rapid economic growth through the 1970s at a period when the economies of northwest Europe were faltering (White 1986). It was still possible to think of the countries of southern Europe as countries of emigration in the longer term.

The late 1970s and the early 1980s, however, started to produce evidence that something new was beginning to occur, and that foreigner populations were starting to increase in these Mediterranean regions. Statistical information was, in fact, lacking on the flows involved, since one of the consequences of the out-migratory past was that few data were collected on immigration in countries such as Spain and Italy, and the population censuses of such countries contained very little detail on immigrant residents. In general, the countries of southern Europe have lacked the administrative systems to identify and to cope with the new sets of migration flows that have, over the past decade or so, transformed their net migrant balances from deficits to surpluses (Castles & Miller 1993, King & Rybaczuk 1993).

By the early 1990s the available statistical information, which always needs interpretation with great caution, indicated the scale of the transformation that had occurred (Table 10.1). In each case the total registered stock of foreigners increased rapidly over the twenty-year period from 1970 to 1990, and has continued to rise since. It must also be borne in mind that the data given exclude those who are not in possession of a legal right of residence and whose clandestine status excludes them from official statistics: these may amount to half as many again as the official figures. Periodic amnesties on clandestine migrants, for example in both Italy and Spain, have resulted in large-scale year-on-year increases in registered foreigners that do not directly relate to actual immigration flows in those years.

Yet it is interesting to note that as late as 1987 the reports produced by SOPEMI, the agency responsible for the collating of international migration data within the Organization for Economic Cooperation and Development, still tabulated the four southern European states shown in Table 10.1 as countries of emigration. Net emigration had actually become net immigration in Greece from 1975 onwards, in Spain from 1979 and in Italy from 1978 (after two other net inflow years in the early 1970s). Only in Portugal was there still net out-movement into the early 1980s.

Table 10.1 Estimates of foreigner population stocks (in thousands), Mediterranean Europe.

	1970	1980	1990
Portugal	23.4	49.3	107.8
Spain	148.4	182.0	277.0
Italy	147.0	298.7	781.1
Greece	93.0 (1971)	182.0 (1981)	229.0

Sources: National Statistical Yearbooks; SOPEMI reports.

The migration turnaround in southern Europe is therefore not a recent phenomenon, but has achieved a degree of maturity over the past ten years or so. It does not simply consist of returnees from earlier emigration episodes, but in a very real sense involves new patterns of migration flow – new both in terms of origin-destination relationships, but also in terms of the relationships of migration with developing economic and political structures. It is worth briefly considering these relationships before proceeding to more detailed examination of the specific cases of Spain and Italy.

ECONOMIC CORRELATES OF MIGRATION

In economic terms, recent and contemporary international migration flows into southern Europe relate strongly to developing post-industrial and post-Fordist trends in European economies as a whole, and to aspects of a new world economic order affecting Third World countries (Mingione 1985, Barsotti & Lecchini 1989). Labour markets have become more fragmented and segmented, with opportunities often being highly localized and specialized in their demands. Migrant labour has been drawn in to various of these niches in southern Europe, and has therefore become widely distributed geographically in ways that do not necessarily reproduce the concentrations of ethnic minorities in large industrial cities that are familiar in northern Europe. In Italy, for example, rural Sicily has been as important a reception and employment area for foreign migrants as have many urban centres in the north (Montanari & Cortese 1993).

Foreign females have been particularly drawn in as domestic service labour, often without proper residence permits, but recruited and shielded by several "welfare" organizations, including the church (Arena 1982, Campani 1989).

Clandestine migrants have found several niches in southern European economies – cleaning car windows in traffic jams, hawking trinkets around tourist resorts, or replacing traditionally more local sources of seasonal agricultural labour (Eaton 1993). There is evidence that, as in other parts of Europe, there is sometimes official tolerance of such migrants since there is a recognition that certain economic sectors have become dependent on their presence.

Migration into southern Europe does not, however, originate entirely within the less-developed world. With the increase in international activity within the global economy, the increased movement of skilled labour around European capitals and other major cities is also a factor increasing the presence of foreigners in southern Europe, with Lisbon, Madrid, Barcelona, Milan, Rome, Athens and certain smaller places all now having distinct, often relatively transitory, communities of skilled workers and their families from other developed countries.

We should also note one type of migration that results from economic wealth, even though it does not involve labour flow. Spain in particular (but other countries to some extent as well) displays the growing significance of retirement migration from northern Europe (Muñoz-Perez & Izquierdo Escribano 1989). This brings a population that is relatively well off, retiring from professional or office employment, seeking retirement in the sun and creating a series of new "foreigner" communities. The connections between such immigration and more traditional large-scale tourist flows are obviously considerable, and it is tourist areas such as the Portuguese Algarve, the Costa del Sol in Spain, and Lake Garda and Tuscan "Chiantishire" in Italy that have seen the most significant growth of foreign property ownership and residence in recent years.

POLITICAL CORRELATES OF MIGRATION

The political dimension of the new international migrant flows into southern Europe can be considered in two different dimensions: the external political forces at work, and the internal political response.

Many commentators have attributed part of the increase in foreigners in countries such as Spain or Italy to the progressive tightening of immigration controls in the traditional northern European destinations (Barsotti & Lecchini 1989, Gozálvez 1990), thus leading to the development of a "buffer state" situation in which Mediterranean Europe absorbs migrants, particularly from Third World countries of the "South", without transmitting them on northwards. Simple geography accentuates this process, since the proximity of North Africa means that the first landfall in Europe for migrants from the South will occur at one of the many ports of that continent's Mediterranean coastline, or at a major international airport such as Rome or Madrid.

As the governments of other states within the European Union seek to implement the policies first suggested in the Dublin agreement, whereby asylum-seekers are to be sent back to the first "safe" country they reached, there is a strong likelihood that the "buffer" role of southern Europe will be further accentuated in future. Asylum-seekers generally increased rapidly in number in southern Europe during the 1980s, although in a highly variable pattern from year to year, reflecting changes in governmental attitudes and periodic amnesties for clandestines: some clandestines claim asylum when threatened with deportation.

In terms of internal political responses, there are signs that public opinion is turning towards more critical attitudes. Racist attitudes are developing in Italy (Campani 1993) and played a role in the strong showing, particularly in the south, of the extreme right-wing Movimento Sociale Italiano (MSI) in the Italian municipal elections of 1993. This party was effectively re-formed for the 1994 national elections as the Alleanza Nazionale, and both it and the Lega Nord are completely against immigration from non-European countries. Following the success of these right-wing parties in the 1994 election, such attitudes are now represented at the centre of Italian political life. Part of the general reaction against immigrants is against what is seen as the inability of the authorities to control the inflow of clandestine migrants – an inability that results from the lack of the need for past controls referred to earlier, and which equally worries the strongly controlling states of northern Europe, creating one of the strongest barriers to the full implementation of the Schengen system and the removal of border controls within the European Union.

The future of migration affecting southern Europe will inevitably depend in part on the circumstances set up during the period of migration turnaround over the past twenty years. However, other factors will also be of great importance – most notably demographic factors in southern Europe and the evolution of labour markets, political responses to immigration and immigrants at a variety of scales, and the evolution of public opinion that will influence the climate within which such political responses are formulated. These possible contextual issues for the future can be more closely examined by focusing on specific national examples.

The future of migration – the Italian case

In considering migration affecting any country it is first necessary to establish certain basic demographic, economic and legislative features of the country, since these are likely to be the most important immediate factors defining future migration scenarios. In the light of these "internal" factors, consideration must also be given to "external" influences such as the policies of other immigration countries, and the demographic situation of the countries of emigration. In the countries under discussion here, future migrant origins may lie in eastern Europe as well as in countries on the southern shores of the Mediterranean.

In terms of its demography, Italy (along with Spain) is distinctive in relation to trends elsewhere, most notably in terms of the extremely rapid fall in fertility – to the level of 1.25 children per woman in 1992 – with average life expectancy gradually increasing to reach 73.2 years for men and 79.7 years for women. This particular combination of fertility and mortality has effectively produced zero population growth in Italy, with a natural growth rate of only 0.02 per cent in 1992. Ignoring any migration effects, the outlook for the near future is of an imminent fall in population size once the effects of the earlier baby boom on the age structure of the Italian population have come to an end: these effects have maintained an enhanced number of births in recent years and have mitigated the impact of mortality (Golini 1994). The most significant effects of fertility and mortality changes occur in the population age structure. In Italy the constant and rapid decline in births has profoundly affected the balance between the age groups, with the proportion of young people gradually declining, causing a growing ageing of the population. Even were fertility to increase again in the future, the speed of this ageing will increase.

Particular attention must be paid to the active population, aged 20–59 years, since there are important links between this age-group and the more general economic and social system (Table 10.2). This age-group currently represents just over 55 per cent of the Italian population, representing a rise from earlier levels, and is large enough to reduce the effects of the fall in births and to absorb the impact of population ageing on the pension and social security system. However, over the next twenty years the rate of increase of the active population will slow down considerably and be reversed as the effects of reduced fertility begin to be felt.

The effects on the labour market of the gradual and later inevitable thinning of numbers in this active age-group will bring important repercussions for the forecasting of migratory flows towards Italy. If in the future there is a reduction in the supply of labour, new flows of immigration could be encouraged. However, this is only one of the possible outcomes of such a reduction since there is currently a considerable amount of slack in the Italian labour market (especially relating to young people and women), and it is also possible to envisage that technological

Table 10.2 Population and age structure in Italy: actual change 1970–1990 and expected change 1990–2010 (%).

Age groups	Actual changes				Expected changes		
	% Total 1970	1970–80	1980–90	% Total 1990	1990–2000	2000–2010	% Total 2010
0–19	31.7	–1.4	–6.0	24.4	–4.1	–1.2	19.2
20–59	52.1	+0.5	+2.9	55.6	+1.3	–1.5	55.3
60	16.1	+0.9	+3.0	20.0	+2.8	+2.7	25.5

Source: Golini (1994). The expected changes relate to the low variant of UN-based projections.

innovations could increase productivity so that recourse to a foreign labour force may not in fact be necessary (Gesano & Heins 1994).

It is possible to identify at least two features of the Italian labour market that might activate immigration flows: first, shortages in domestic labour supply; and secondly a generalized readiness on the part of Italian firms to hire non-European Union workers on terms below those dictated by the market. At present labour shortages only affect regions in the centre-north of Italy and are limited to certain jobs. However, there will probably be labour shortages for less qualified and heavy manual work in the relatively near future. Over recent years immigrants have found jobs most easily in agriculture and fisheries and in the service sector (including street vending, domestic help, porterage and catering): they have only recently begun to work in industry, but numbers are increasing, especially in construction and in public works.

We can thus see the potential importance of migration and its determinant role in the near future in both strictly demographic terms (as a potential counterweight to changes caused by the incipient decline in the indigenous population), and in economic and social terms in relation to labour market dynamics and to the processes of evolution and integration affecting foreigner communities.

As was pointed out in the previous section of this chapter, Italy has only undergone foreigner immigration fairly recently. Figure 10.1 indicates the rate of growth in the total number of residence permits for foreigners since 1970. Besides the consolidation of migration flows into Italy originating in Africa and Asia, it is also important today to take into account the new mobility made possible by the profound political changes under way in the whole of eastern Europe. Such flows could have an enormous impact on already fragile social cohesion within various western European countries, and the position in Italy would be particularly problematic, both because of contemporary political upheaval in the country and because of the potentially large flows resulting from Italy's geographical proximity to certain countries of origin, as was seen in the flight of Albanians to Apulia in the summer of 1991.

167

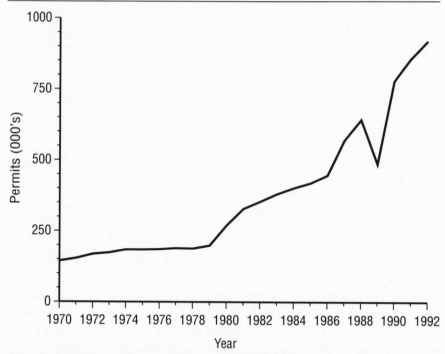

Figure 10.1   Total residence permits in operation for foreigners, Italy 1970–92.
Source: updated from Birindelli (1990).

A picture of the current situation of foreign workers in the Italian labour market can be gained by considering the latest available data on foreigners in Italy, published by the national statistical office (ISTAT). These data, relating to 31 December 1992, show that, considering all legal migrants together, work is the main motivation. Non-labour migrations are nevertheless fairly important, representing 44.1 per cent of all migrants. Table 10.3 highlights the fact that foreigners coming to Italy from Africa or Asia generally did so with the intention of finding work, such that migration from countries outside the European Union or from the Americas involves a first generation of labour migrants with very few movements involving other family members as yet. However, between 1990 and 1992 there were 22000 requests for family reunifications, of which 13000 were permitted, chiefly for family members from the Third World and from eastern Europe. This kind of migration is likely to increase in the future, given the size of the preceding labour flows. By contrast, migration from western Europe, and particularly from the other side of the Atlantic, has much less relation to the labour market.

One other migrant component that could affect the short-term development of non-labour movement concerns refugees. Requests for political asylum exceeded 26000 in 1991, but only 800 applications were granted – giving Italy one of the highest rejection rates in Europe. At the end of 1991 Italy sheltered 12103 refugees,

Table 10.3 Residence permits according to reason for application, by main areas of migrant origin (31 December 1992).

Area of origin	Total permits (000s)	Principal reason for application (%)		
		Work	Family	Other
Europe	325.9	42.3	14.9	42.8
EU	146.8	39.5	17.3	43.2
Non-EU	179.1	45.3	13.2	41.5
Africa	283.8	77.8	5.8	16.4
Mediterranean	175.9	81.5	6.3	12.2
Asia	158.5	66.0	10.3	23.7
Americas	148.7	28.5	30.6	40.9
Latin America	81.9	35.3	20.4	44.3
Total	923.6	55.9	13.9	30.2

Source: ISTAT (1993). *Rapporto annuale 1992. La situazione del paese.* Rome.

of whom 42.9 per cent came from Europe: a further 17000 refugees from the former Yugoslavia were accepted as a United Nations quota in August 1992.

However, official data only provide a partial picture of the size and origin of migration flows. Figures for immigration from the Maghreb, for example, are underestimates because there are large numbers of illegal immigrants from these countries so that the number of residence permits issued does not reflect the true situation. Other sources of official data, particularly on "regularizations", confirm this problem. Italy passed amnesty laws in 1986 and 1990 allowing illegal immigrants to register with the authorities to become "legal". In 1986, 118709 registered in this way, and in 1990 a further 234786 foreigners did so: such numbers illustrate the problems of data interpretation and comparability (Bonifazi 1993).

To understand the migration issue, and above all to put forward hypotheses about possible future migration scenarios, it is essential to look at political attitudes and relevant legislative frameworks in Italy and to evaluate the extent to which these are in line with immigration policies in other host countries – particularly within the European Union. The comparative vulnerability of Italy to uncontrolled immigration flows has been alluded to in the introduction to this chapter: in particular, Italy has borders (especially coastlines) that are difficult to control, and is also proximate to potential sending areas such as Albania and the ex-Yugoslav republics.

Italy's current legislation on foreigners from outside the European Union was passed in 1990 and does not differ from the approaches adopted by other European countries. The main objectives of the law involve the control and regulation of admissions and of foreigner residents, with particular attention being paid to illegal immigration. Straightforward entry to Italy is only granted to certain categories of applicant, such as refugees, the families of legal immigrants, or workers with the

169

qualifications necessary to fill particular gaps in the Italian labour market. Legal immigrants enjoy civil and social but not political rights. The legislation is designed to help immigrants to settle in and integrate, at least in theory. One original element is a provision for the annual programming of immigration flows, which effectively recognizes the possibility of encouraging such flows if and when the labour market can absorb them (Bonifazi & Gesano 1994).

The need to accept and integrate foreign minorities has become a problem in all advanced European countries, but the exact details of these objectives vary according to several factors: the length of time the flows have been in operation; their size; their composition; the relevant legislative and institutional frameworks; and the economic and social organization of the host country. It is in these problematic areas of policies for receiving and integrating foreigners that there are differences between the more recent European immigration countries such as Italy and those where immigration is already consolidated. For these latter countries (such as those of northwest Europe) immigrant, or more properly "ethnic minority", communities pose what might be called "second stage" problems concerning their integration into the civil and social fabric of their adopted country, having already gone through the initial stages of arrival, finding a job, and settling in.

In contrast with such situations, the contemporary situation in Italy is particularly complex. Several problems can be seen to exist simultaneously. First, since immigration is a very recent phenomenon, the most urgent problems concern arrival and reception. Among the problems that are still unresolved for many immigrants in Italy are those concerned with registration with the authorities, finding work, housing, and access to welfare services. Planned intervention to ease these problems is essential. The experience of other countries, and the expectations of communities who are already resident in Italy, indicate that learning the language, access to education, opportunities for training, and the acquisition of professional qualifications are the most important ways of helping immigrants to achieve smooth integration into the life of the host country. In fact, Italian laws on these various themes provide for the equality of treatment and rights for non-EU, EU and Italian workers, along with the right of access to social and health services, schools, and housing, and the right to maintain one's cultural identity.

However, as mentioned earlier, the overall situation in Italy is one in which there is a serious risk of growing intolerance towards foreigners. Some of the reasons for this include the complex overlapping nature of the responsible authorities (including institutional bodies at central and local level – the regions as well as the communes – as well as both public and private bodies, such as voluntary organizations and the Church); the particular concentrations of foreigners in some areas of the country; and the pre-existing and unresolved internal questions of the Italian political economy such as the North/South divide in Italian life. Immigrant issues have therefore been added to the maelstrom that currently constitutes the Italian political scene, and can be used by different political factions in ways that are perceived by the parties to bring the greatest advantage to their own interests.

Forecasting migration flows into the future is an extremely complex exercise. In contrast with the other demographic components of population development such as fertility and mortality, which have lower variations and generally medium or long-term effects, migration is subject to very intense and sudden fluctuations that may depend on political and economic events that are difficult to foresee. For this reason, demographic forecasts consider the migration component with great caution. United Nations' projections, for example, only take the migration variable into account for a few industrialized countries, whereas the estimates produced by individual governmental organizations in various countries are for the most part based on simple hypotheses such as the constancy of current levels of movement, a growth towards potential or desired levels, or decline (Wils 1991). Rather than forecasting flows in numerical terms, the goal must be to highlight the significance of different issues such as impacts on the overall size of the population, on ageing processes, on social security systems, and on the cost of public services.

One particularly useful strategy is to prepare alternative scenarios on which to base political, economic and other choices. The choice of hypotheses for the construction of such scenarios is a delicate issue. Alongside traditional methods of estimating entry flows it is becoming increasingly important to evaluate the reproductive behaviour of immigrant populations now established in the host countries. The concentration of immigrants in the reproductive age groups, and the high (but declining) fertility models that characterize them, are two important elements in the evaluation of the full impacts of immigration on the age structures in the host countries. In particular, the effect of integration processes followed by immigrant groups, measured as their degree of adaptation to the demographic behaviour of the host population, appears to be of particular significance, above all in the European context (Lutz & Prinz 1992, 1993).

There are no official forecasts of immigration flows for Italy, but there are several studies that analyze specific aspects of such migration. What follows here are derived from two different scenarios developed in such studies: the first deals with the effects of migration flows on public spending, and the second with the effects on demographic structures. The time period under consideration differs between the two studies: the first considers economic impacts over a short period up to the year 2002, whereas the second makes forecasts for a much longer period of up to 100 years in order to evaluate more fully the changes in demographic variables.

Estimates of the numbers of foreigners and the structure of the foreign population in Italy by 2002 have been made by the Brodolini Foundation (Fondazione G. Brodolini 1992). The method used was to take the situation in France in the first half of the 1980s as a reference point, since it was considered that similar conditions could be reasonably expected to develop in Italy over the next ten years. In partic-

171

ular, certain basic elements of the French situation are similar to those in Italy, such as the geographical locations of major sending areas, and various legal aspects of reception (such as nationality legislation and the recognition of family ties). The study produced three different scenarios (Table 10.4), all showing a considerable migration impact. Even the most moderate scenario (variant A) implies the failure of the flow containment policies that the Italian government has adopted since 1990 to try to stop the trends that had developed over the previous decade. It should be noted that the study deliberately produced a high migrant population variant in order to estimate public spending costs in the most critical case, but recent trends have suggested that such a high variant is not, in fact, completely unrealistic. Even the lowest change scenario in Table 10.4 produces a higher estimate for foreigners than was being produced by forecasts in the mid-1980s (Collicelli & di Cori 1986).

Table 10.4 Hypothetical situation in Italy, 2002.

	Scenario A	Scenario B	Scenario C
Total population (000s)	59313	60199	61647
Italian population (000s)	57455	57455	57455
Foreign population (000s)	1858	2744	4192
Foreigners/total population (%)	3.13	4.50	6.80
Annual increase of foreigners 1990–2002*	73,789	147,579	268,265

* Foreigner population of Italy in 1990 taken as 972804.
Source: Fondazione G. Brodolini (1992).

The demographic aspects of the development of fertility and migration are the subject of the second hypothetical study (Golini 1994). This was carried out to examine the effects of the marked fall in fertility on the age structure of the population of Italy and considers three different variants for future fertility evolution and for immigration flows (Table 10.5). Summarizing the results of a complex study, there is a suggestion that significant demographic change in Italy will only start to appear after the period 2005–10, but will become very evident after 2025–30 – within the lifetimes of a majority of those alive today. The problems of the overall size of the Italian population could be solved by encouraging large immigration flows (the results given in Table 10.5 relate only to a relatively modest inflow of 50000 per annum), but that in order for problems of population ageing to be reduced these immigrants would have to import and retain their high fertility reproduction models. The effects of this on the cultural identity of the population of Italy were outside the aims of the study. Given the fact that Italian fertility is now below the 1.30 level used in the simulation, it is clear that were this level to be maintained overall, the long-term scenario, without massive immigration, is a drastic reduction in the Italian population to a figure less than half that of the present day.

Table 10.5 Effects of low fertility and immigration on variations in the Italian population by age group, 1992–2092.

	With no immigration				With 50000 immigrants per year			
	0–19	20–59	60+	Total	0–19	20–59	60+	Total
Population 1992 (million)	13.2	31.6	11.9	56.7	13.2	31.6	11.9	56.7
Average no. children per woman	Total variation in millions, 1992–2002							
2.07	–	−5.5	+2.5	−3.0	+2.2	−1.0	+4.3	+5.5
1.80	−5.1	−13.0	+0.2	−17.9	−3.3	−9.1	+1.8	−10.6
1.30	−10.5	−23.0	−3.8	−37.3	−9.5	−20.0	−2.3	−31.7
1.00	−12.0	−26.9	−5.9	−44.8	−11.3	−24.3	−4.5	−40.1
	Average annual percentage rate of variation, 1992–2002							
2.07	–	−0.2	+0.2	−0.05	+0.2	−0.03	+0.3	+0.1
1.80	−0.5	−0.5	–	−0.4	−0.3	−0.3	+0.1	−0.2
1.30	−1.6	−1.3	−0.4	−1.1	−1.3	−1.0	−0.2	−0.8
1.00	−2.4	−1.9	−0.7	−1.6	−2.0	−1.4	−0.5	−1.2

Source: Golini (1994).

The situation of Italy in the face of possible future migration trends is an interesting one, compounded of the recent emergence of the country as a major migrant destination with the contemporaneous evolution of Italian fertility to its lowest ever level. Assuming that, in the near future, fertility and mortality rates will not change significantly and that migratory pressure from abroad will probably increase, future scenarios will be very much influenced by two forces: the labour force effects of the further growth of a middle class made up of Italian natives, which should result in vacancies in those low-level jobs that Italians do not want to do because of low pay or social undesirability; and the measures taken by the government with respect to the integration of non-EU immigrants, and in particular steps concerning the granting of Italian nationality and the rights of immigrants for family reunification.

The future of migration – the Spanish case

In certain respects the migration turnaround in Spain, although involving a smaller total number of in-migrants, has been even more remarkable than in Italy. In Italy, with its "North–South divide", there had been some diversion of migration streams originating in the south away from international destinations and towards the cities of the north and northwest (Turin, Milan and, to a lesser extent, Genoa) for some decades. Certainly in Spain there had been long-standing internal migration to

Barcelona and to Madrid (Estébanez & Puyol 1973) but these had not constituted true industrial regions in the same sense as Lombardia or Piemonte in Italy. When the strong net inflow of foreign migrants started in Spain, during the 1980s, several possible reasons were of relevance. As elsewhere, restrictions on entry into other countries of the then European Community were of importance, but in-migration was also a response to Spain's own high levels of economic growth, especially during the second half of the decade, coupled with the more general stagnation of the economies of other advanced industrial countries. It could be suggested, in fact, that Spain's membership of the European Union (Spain joined only in 1986, along with Portugal: Greece had joined in 1981, whereas Italy had been a founding member of the Community) brought both economic growth and changed international migration flows as interrelated outcomes. Although Spain's fertility rate is joint lowest in the world with Italy's, it is quite clear that migrants are not arriving to cover demographic decline: indeed, despite recent economic growth Spain remains a country with a severe problem of unemployment. It is doubtful whether, given any other current fertility rate, the migration situation would be any different.

It is almost impossible to know exactly how many foreigners there are in Spain at any one date. The latest published statistics on foreign residents with the legal right to be present in the country (dating from December 1992) show a figure of just under 400000 residents. However, estimates of the number of clandestine migrants, to be added to this figure, vary enormously (Izquierdo 1992a, Colectivo Ioé 1993). In 1991 a special regularization process for workers was implemented, which resulted in 132000 applications from clandestine or "illegal" migrants, of which 109000 were granted legal status. The foreign labour force of Spain can be estimated in 1992 as possibly around 212000, a figure arrived at by adding those with legal status at the outset to those legalized through the regularization procedure. There are thus two distinct groups within this foreign labour force (Pumares 1993): one group contains those who entered with work permits, and the other those who entered illegally and who later underwent regularization (Table 10.6). Europeans are predominant in the first group, particularly European Union citizens including large numbers from Germany, the United Kingdom and France. The Portuguese, actually the largest European group, in fact play a different role in the economy from that of other European Union citizens, and are more similar in this respect to those originating in the developing world. Other significant groups amongst the regular foreign work force originated in Morocco, Argentina and various other parts of the Americas.

However, the majority of Moroccans were present in the Spanish labour force only as clandestine workers who underwent regularization in 1991: indeed Moroccans made up 45 per cent of those whose residence became legal as a result of that procedure, with migrants from Latin America following (López 1993). The number of European Union citizens regularized in 1991 was insignificant, although

Table 10.6 Foreigner labour force in Spain, December 1991.

	Work permits			
	Under normal procedures		Under the regularization process	
Countries of origin	Total	%	Total	%
Europe	50692	48.8	8187	7.5
EU	43 955	42.3	2525	2.3
Germany	8423		209	
Belgium/Lux.	1398		45	
Denmark	842		24	
France	5977		191	
Greece	151		12	
Ireland	856		50	
Italy	3631		212	
Netherlands	2523		84	
Portugal	10430		1178	
UK	9724		520	
Rest of Europe	6737	6.5	5662	5.2
Africa	14735	14.2	59983	55.4
Morocco	10219		48233	
Rest of Africa	4516		11750	
North America	5236	5.0	1576	1.5
USA	4061		1047	
Rest of N. America	1175		529	
Central and S. America	19813	19.1	28488	26.3
Argentina	7705		7404	
Chile	1814		2328	
Peru	1195		5664	
Dominican Rep.	877		5517	
Rest of C. and S. America	8222		7575	
Asia	13025	12.5	9902	9.1
China	1935		4219	
Philippines	4694		2609	
Japan	1333		229	
Rest of Asia	5063		2845	
Total	103884		108308	

Source: Ministry of Labour (unpublished data).

over 5000 eastern European migrants, most of whom were Polish, were dealt with by the procedure. Together these eastern Europeans, along with the Africans and Asians, displayed the fastest growth rates.

Regional distributions of foreigners do not vary according to origin. All migrants are concentrated in the Mediterranean coastal provinces, in Madrid and Barcelona, and in the two groups of islands (the Balearics and the Canaries). These are the most dynamic areas of Spain in terms of their economic structures, and together they contain between 70 and 80 per cent of each migrant group.

Discussion here so far has concentrated on the foreign labour force. To this must be added dependants, both those legally resident and those clandestine. Although males predominate in the total migration streams (counting both labour migrants and their dependants), as would normally be expected in a relatively new international labour migration situation, the percentage of women is not negligible at around 40 per cent. As in Italy, this results in part from demands for specifically female labour for domestic service work, as well as from some family reunification.

However, age and sex distributions are not uniform for all nationalities. Populations over the age of 65 are locally important in certain tourist regions along the Mediterranean coast and are made up of more than 80 per cent Europeans. Women represent over half of all those of Latin American or European Union citizenship living in Spain, whereas amongst Africans they do not even reach 30 per cent.

The second generation is as yet small, at around 8 per cent of the foreign population, and more than 55 per cent are the children of Latin American or European Union citizens. However, this is a rapidly changing phenomenon as the Moroccan community (the largest) will increase significantly throughout the near future: already 20 per cent of foreign children under the age of two are the offspring of Moroccan parents. It should be noted that three-quarters of the Moroccans living in Spain were subject to the legalization process of 1991, and since then the demands for family reunification have been very strong.

Children of foreign parents have full access to public education, but those over the age of 12 who arrive without being able to speak Spanish find many obstacles in entering the education system. The concentration of such children in some districts of Madrid and Barcelona has forced the rapid introduction of changes in schools in order to meet new demands. There is some evidence of poorer school performance among Moroccan children, including the early abandonment of their studies, which will produce future restrictions on the social and occupational mobility of this group.

The employment structures of the various immigrant groups are significantly different (Figs 10.2 and 10.3). Among those following legal entry and work permit procedures a much higher proportion were employed as highly skilled workers (professionals and managers), whereas amongst those who entered the labour force illegally there was an overwhelming proportion employed in unskilled jobs.

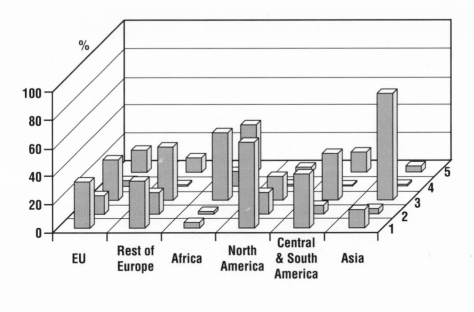

1   Professionals/Managers
2   Administrative
3   Sales/Sevices personnel
4   Agricultural workers
5   Industry/Building workers

Figure 10.2   Foreign labour force in Spain: those with normal work permits, 1991. *Source:* Ministry of Labour (unpublished data).

Taking first the case of those with regular work permits (Fig. 10.2), it must be noted that the category "professionals and managers" is a relatively heterogeneous one, including high-level professionals along with high and medium-level technicians and managers. There is a clear pattern whereby those classified in this group originate in advanced western countries (particularly from the rest of the European Union, and the USA), whereas many of the relatively few Japanese are also in this category (Rodríguez 1993). This is clearly the result of the internationalization of the labour market for such employment, and the activities of multinational firms. Amongst the "legal" immigrants Latin Americans make up a labour force of medium skill levels, whereas Asians and Africans tend towards domestic and personal service and trading, which are low-skill tertiary activities. Foreign agricultural and building workers are mostly African.

Amongst those who entered the Spanish labour market illegally the information obtained from the regularization procedure shows that their insertion into employment was both different and complementary (Fig. 10.3). The key variable in the distribution of employment among regularized immigrants was, once again, their

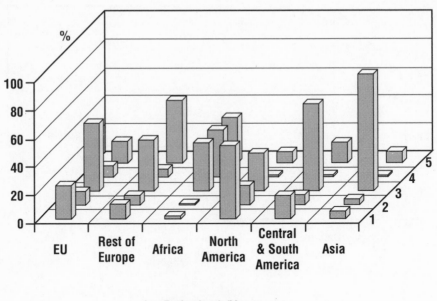

Figure 10.3 Foreign labour force in Spain: work permits issued via the regularization of clandestine migrants, 1991–2. *Source:* Ministry of Labour (unpublished data).

origin. Most such migrants were involved in low-level tertiary activities (particularly petty trading and domestic service), and this was particularly so amongst Asians and Latin Americans (80% and 90% of the clandestine migrants of these origins respectively). The presence of Africans (mostly Moroccan) among those involved in agriculture (32% of regularized Africans) and construction work shows that they are concentrated in jobs that are unattractive to the Spanish population, confirming their role as substitutes for Spanish workers. There are signs that the same is becoming true for migrants from eastern Europe.

In summary, the government-inspired regularization of clandestine migrants answered a need to provide a legal status and response for a flow of workers, families and refugees – a flow that resulted from innumerable individual migration projects that converged on Spain because of its image as a developed country with economic opportunities, and with easy access and entry possibilities. On the other hand, those who enter Spain through the obtaining of normal work permits are using the Spanish labour market as an extension of their own socio-economic environment.

SPANIARDS' ATTITUDES TOWARDS IMMIGRATION

Several surveys have recently been carried out enquiring into attitudinal responses to immigration in Spain and elsewhere. The overall picture that emerges from these surveys is that most Spaniards do not express particularly negative attitudes towards foreign immigration. This is especially noticeable when comparisons are made with attitudes in other European Union countries. The 1992 Eurobarometer survey lists Spain as the least racist country in the analysis, a fact that is undoubtedly related to the relatively few foreigners there.

The surveys show a strong rejection of what might be called more obviously racist attitudes. A survey carried out for the Spanish Centre for Sociological Research in 1993 showed that 83 per cent of those questioned expressed their total disapproval of "openly discriminatory acts" relating to religion or nationality, and 80 per cent said they would never vote for a racist party, as opposed to the 7 per cent who would do so in certain circumstances. Nor were respondents very strict as regards entry restrictions, although significant differences emerge depending on whether the question is a hypothetical one or whether it specifies a more concrete and immediate situation: thus 84 per cent agreed that everybody should be able to settle in any country, but 67 per cent were in favour of entry quotas for Spain.

The surveys also showed favourable attitudes towards immigrants being given access to public health care or education (over 80% support), and even towards the provision of help to find somewhere to live (70%). Respondents also claimed to pay little attention to whether or not they had immigrant neighbours. However, on these housing issues the translation of attitude into practice is doubtful since, on the one hand, immigrants find there is much reluctance amongst landlords to let property to them and, on the other hand, whenever steps have been taken by various authorities and agencies to provide immigrants with accommodation or to improve their conditions opposition has arisen amongst neighbours. The most negative attitudinal responses in the surveys relate to unemployment and, to a lesser extent, crime: according to another survey carried out in 1992, 55 per cent of respondents thought that unemployment was rising because of immigration, whereas 44 per cent believed in a link between immigration and rising crime.

Looking at the survey results in total, it might be said that the points concerning foreigners that arouse most controversy among Spaniards are, in descending order, jobs, crime and housing. Jobs and housing are spheres of life where problems can be considered as arising from competition (Izquierdo 1992b). As the unemployment rate in 1993 was over 20 per cent and housing policies have clearly been insufficient to cater for rising demand over the past few years, both these problems are close to many Spaniards. Although health care and education are guaranteed for the whole population, jobs and housing are both scarce for Spaniards themselves and are thus more sensitive issues. However, 71 per cent of respondents acknowledge that immigrants do jobs that Spaniards do not want to do. It is, nevertheless, around the question of employment that opposition to immigration tends to crystallize, with the argument of unemployment effects being used by those who actually oppose immigration for other reasons but who can bring their opinions closer to those of more moderate elements by relating immigration to a subject that also concerns them. This is a tactic that has been very successfully used in France with Le Pen's simple equation linking the number of jobless with the number of immigrants – a tactic that ignores the realities of labour force structures.

Perceptions of crime and lack of safety emerge strongly as concerns, both in the surveys and in other research based on discussion groups (Pumares & Barroso 1993). These concerns appear to relate to two issues. First, there is a specific trend of thought according to which immigrants from less developed countries, because they are poor and have difficulties finding work, "naturally" turn to crime. Secondly, because some of these immigrants are highly visible in certain symbolic areas of central Madrid and elsewhere, sometimes performing illicit or less reputable activities, this image is extended by the general public to all immigrants, including the majority who are "unseen".

However, there are several indications from the surveys and other research that we are still at an incipient stage in the formation of Spanish public opinion towards immigration and immigrants. Many people have not yet defined their own position. This is, of course, a reflection of the fact that the number of immigrants is still relatively low such that the majority of the Spanish population has not really been affected by them. What affects people most are the "visible" groups seen in city centres and these are therefore problematized. Although the lack of direct experience of immigrants leads many Spaniards to continue to voice opinions of solidarity with immigrants, such opinions may change. In areas with high concentrations of immigrants, social conflicts have arisen that point towards caution in any optimistic predictions about the maintenance of Spain's position as a non-racist society.

Nevertheless, by late 1993 Spanish public opinion was less exercised about immigration than it had been a year earlier. For almost two years, throughout 1991 and 1992, there had been daily press reports concerning illegal immigrants, the regularization process, and Moroccan immigrants landing by small boats on the Andalusian coast: these reports have now largely stopped. The current economic crisis has considerably lessened the flow of immigration, but public opinion does not seem to have turned strongly against the immigrants to use them as scapegoats. Surveys taken in 1993 show more tolerant attitudes than those of 1992. It could be argued that Spain is entering a transition period during which Spaniards will begin to accept that they are living in a country of immigration.

IMMIGRATION POLICIES AND THE FUTURE OUTLOOK

Recognition of the role of immigration in Spain has led towards a need for real policies. In 1985 Parliament passed a Law on the Rights and Liberties of Foreigners in Spain, which unified the hitherto dispersed regulations on such individuals. The law has been controversial and has been considered as being restrictive in certain quarters, but it was not put to specific political use until 1991 when the general outlines of a Spanish immigration policy were laid out in the "eleven points" approved by Parliament (Comunicación del Gobierno 1991). These may be summarized as follows:

- control on the flow of immigration, cutting off clandestine entry as much as possible and setting up a quota system
- taking action against clandestine employment by intensifying inspections of employers and by strengthening the available sanctions
- reducing crime by increasing police vigilance and encouraging judges to deport foreigners who commit petty crimes
- improving the administrative structures for handling immigration and immigrant issues, which, as mentioned at the start of this chapter, are not well adapted for the contemporary situation
- taking steps towards social integration and improved access to education, health care and housing.

However, these policy objectives have received unequal attention over the past few years. Arguably point 1 above (dealing with immigration control) has been the most effectively implemented, mainly through increased frontier controls. Visa requirements have been imposed on residents of North African countries, and immigration service vigilance along the Andalusian coast has been increased against the arrival of Moroccans in small boats. At the same time an agreement has been reached whereby the Moroccan government is now seeking to control departures from its own shores. In 1993 the first quotas on foreign workers were established, although they were not immediately put into effect – a comment on the administrative efficiency referred to above. A total of 20 600 workers were to be admitted during the year, including 10 000 in agriculture and 6000 in domestic service. Just over half the total were to be temporary workers, and the leading regions for quota entries were to be Catalunya (24%), Madrid (14%) and Murcia (12%). One problem in implementing quotas is that Spanish entrepreneurs need to become accustomed to using such a system. Their position is made more complex by the fact that the economic system in which quotas are to be imposed is a very different one from that which existed elsewhere in Europe during the period of regulated migration in the 1960s.

In those days a large part of the demand for foreign labour came from large companies; today in Spain the demand for foreign workers comes from small firms, sometimes as small as a single household, who are looking for the most flexible workers to cover immediate needs. It is practically impossible for such firms to foresee how many workers they will need over the coming year, and for how long, apart from the fact that the firms will have to lose part of their autonomy and flexibility in the use of foreign labour by hiring through official organizations. Contemporary post-industrial systems of flexible accumulation are, as suggested earlier, intimately connected with labour supply issues.

Other of the government's "eleven points" have been implemented to a greater or lesser extent. As regards crime (item 3 above), recommendations to judges that they should issue expulsion orders on foreigners who commit minor offences have had little significance, and the measure is still only being used against foreigners residing illegally in Spain. Work inspections (point 2 above) have not notably increased. A single office for handling permits has been created (these previously had to be processed by both the Ministry of Labour and the Ministry of the Interior) although the present economic crisis has meant that it is inadequately staffed. There is still a lack of definition of the duties of each administrative department towards immigrants.

Most of the integration measures have been suspended up to now, but recently it has become the intention that family reunification regulations should be modified, and it will also be made easier for foreigners born in Spain (or who arrived at a young age) to obtain Spanish nationality. Attempts to set up immigrant housing schemes are in a state of paralysis. However, schooling for all children is guaranteed, as also are health care and unemployment benefits for those who make their social security contributions.

The short-term outlook for future immigration in Spain depends to a great extent on the country's economic progress and the development of the labour market. Large-scale recession throughout the western world over the past few years has been leading to the reconsideration of inflexibilities in the labour market, particularly in Spain where regulations are relatively rigid. If the flexibility of the Spanish job market does increase, perhaps with a reduction of unemployment benefits in real terms, then Spaniards may return to some jobs that they had begun to abandon, and this may affect the job supply for foreigners, many of whom are to be found in these activities.

So far this has not happened. The economic crisis has had a drastic effect on unemployment, both for native Spaniards and for foreigners, but it has not brought any major change to the structure of the job market. Those foreigners who are less qualified and who have not yet found a stable job despite becoming legal residents (almost half of such foreigners), have undergone longer periods of inactivity: in order to maximize flexibility some companies have acted to make jobs more precarious, with short-term lay-offs and manipulation of wage rates to disguise real cuts. However, despite the deterioration in the conditions of employment many people have managed to survive by accepting these new circumstances and taking on extra work whenever it is available.

The flow of immigration from the less developed world would appear to have abated because of the decrease in labour demand in Spain and because of tighter frontier control, but there is without doubt a latent pressure for further movement, which may increase again once the economic situation improves. Despite controls there is a clear current tendency for "de facto" family reunification outside the administrative procedures. It seems particularly difficult to control the southern Mediterranean Sea border with Morocco, and there will be a periodic need for renegotiation of the agreement that Morocco should also police this frontier. The quota policy on immigration flows looks destined for failure, given the characteristics of the companies that will need labour.

Nevertheless, although it has not been highlighted in this section on Spain, we should also note that by the early 1990s Spain, like Italy, had arrived at the position of an extremely low total fertility rate, such that the population of the country will cease to grow within the medium term, and will undergo considerable changes in its age structure. The contribution of immigration, and of immigrants already in Spain, to the country's demographic as well as economic future will be important.

Discussion

These two case studies of Italy and Spain have deliberately highlighted different dimensions of the current evolution of immigration. Demographic change, labour force evolution, economic restructuring, public opinion and governmental initiatives will all play vital roles in the future development of immigration scenarios. Throughout southern Europe economic growth at the turn of the twenty-first century will be based on smaller production units than were operative during the great period of migration to northwest Europe in the early post-war years, and in this post-industrial situation of diffused economic growth the relations between economic change and immigration will be complex and two-way in nature. The old models developed in northwest Europe will not be relevant, either to the actual migration process or to the development of ethnic minority communities out of family reunification and immigrant fertility.

Renewed economic growth in the European economy as a whole will almost certainly accentuate income differentials in the Mediterranean geopolitical realm. A recent econometric study (Goldin et al. 1993) has suggested that the effects of the liberalization of world trade to be brought about by the GATT agreement reached in December 1993 will be beneficial in the European Union but negative in the Maghreb states of North Africa and in the other areas around the Mediterranean. If there were to be a further move to completely free trade the discrepancy would be yet wider with both the Maghreb and the rest of the Mediterranean Basin experiencing real GDP declines of 2.3 and 2.4 per cent by 2002 as a result of liberalization alone, whereas in the European Union there would be a GDP growth of 2.8 per cent for the same reason.

These estimates, although interesting, possibly overstate the differential somewhat since southern Europe's continued level of dependence on agriculture means that trade liberalization, with its negative effects on farm prices in the European Union, will not necessarily be an unmixed blessing. Nevertheless, whatever the detail of the figures the outlook is reasonably certain and that is that the income differentials across the Mediterranean will widen in at least the medium term.

Will the Mediterranean become a new Rio Grande? Will the nightly flow of "wetbacks" from Mexico into the USA, and the very considerable adjustments that are occurring in the labour market and, increasingly, the demography of the "frontline states" of the USA be reproduced in southern Europe? The Rio Grande has been the clearest and best defined border between an advanced industrial and a developing world society (House 1982). The Mediterranean and its associated seas make up a wider barrier to cross but one that is not insuperable, as the Moroccan arrivals on the Spanish beaches of the Straits of Gibraltar, or the flotilla of Albanians arriving in southern Italy in 1991 testify. Many parts of southern Europe will still rank for years to come as some of Europe's poorer regions, but there exist opportunities for immigrants there, partly in replacing departed native labour, and partly through the employment possibilities in economies moving towards a consumption function as their basis.

The economic differential will be accentuated, as the discussion of Italy earlier in this chapter indicates, by demographic developments. Not only do the countries of southern Europe now display some of the lowest fertility rates ever recorded (Spain and Italy with total fertility rates of below 1.3; Greece and Portugal on 1.5 – all well below current levels in, for example, France or the United Kingdom), but they also face across the Mediterranean to countries with some of the fastest contemporary rates of population increase. With a rise in Islamic fundamentalism, currently occurring in Algeria, it is even possible to envisage fertility rates increasing slightly in the Maghreb into the future. In 1991 the total fertility rates of Morocco, Algeria, Egypt and Turkey were 4.5, 5.4, 4.5 and 3.7 respectively. In the first three of these, population totals are expected to grow by at least 50 per cent between 1991 and 2010, according to United Nations projections. With real declines in economic wealth and massive increases in populations, the pressures for emigration across the Mediterranean will be considerable. A further significant factor will arguably be the growing cultural hegemony of European models, accentuating the desire, especially among the young, for migration to the "bright lights" north of the Mediterranean (Golini et al. 1993).

The most hazardous element of futures prediction in migration must lie with refugee flows. However, some tentative suggestions can be made. First it is notable that both Italy and Greece are geographically positioned in such a way that they could receive a considerable influx of asylum seekers from escalating conflicts in eastern Europe. By late 1993 Italy was providing shelter for over 20 000 refugees from Bosnia and from earlier fighting in Croatia. Greece was much less affected, but the potential for large-scale population movements there is considerable, particularly if ethnic conflict breaks out involving the Greek minority population of southern Albania. We should also remember that Greece has been involved for some years in one of the less publicized East–West migration flows – the repatriation of peoples of Greek ethnicity from the Black Sea shores of the former Soviet Union.

Italy, with its border with former Yugoslavia, is particularly vulnerable to refugee flows from both East and South. Spain has the potential, given the shared language, to act as a place of refuge for those fleeing persecution in various parts of Latin America. There is a certain possibility of attempted flight from unrest in Algeria and other Arab states within at least the short term, although during the summer of 1994 the government of France made public declarations that it would not consider asylum applications from those fleeing Islamic fundamentalism in Algeria.

Although these refugee flows would all impact directly on southern Europe, a further influence of asylum seekers could occur through a re-direction of such flows towards countries that are perceived as offering greater opportunities for entry. During 1993 many more traditional refugee destinations in northern Europe, amongst them Germany, the Netherlands, France and the United Kingdom, acted to restrict asylum applications and to reduce the likelihood of those not granted refugee status staying on with various forms of exceptional leave to remain. In contrast, as has been made clear earlier in this chapter, administrative systems dealing with immigration into southern European countries remain relatively weak and many potential asylum applicants may enter as clandestine immigrants.

It is not plausible to predict future developments in migration solely in terms of projections forward from current trends: to do so for other demographic indices would yield ludicrous results – for example to project the recent decline in total fertility rates in Italy or Spain would yield zero rates early in the twenty-first century. The influences on migration will undoubtedly change. Demographic and economic developments have already been considered, along with the possible effects of political change elsewhere. Internal political issues will also be of significance. As this chapter has demonstrated, whereas overall public opinion in Spain may still be equivocal about migration, some of the concerns being expressed are redolent of issues that led debates in northern Europe leading to the imposition of strict controls on immigration twenty years ago. Political parties adopting anti-immigrant standpoints are starting to emerge in southern Europe, particularly in Italy. Racist attacks on immigrants are commonplace in Germany, France, the United Kingdom and elsewhere, and there are occasional but increasing reports of such events starting to occur south of the Alps and the Pyrenees. As public opinion becomes aware of the scale of immigration it is quite plausible to see a new social climate arising in which the institutional racism characteristic of many more traditional receiving countries is reproduced in a new Mediterranean context. Although the economic dimensions of movement may be somewhat different from those obtaining in northwest Europe in the early post-war period, the immigrants on whom the focus has been maintained throughout this chapter nevertheless fulfil many of the same general roles – as low-skill workers in socially undesirable occupations that become increasingly stigmatized as a result of the association with immigrants. The Spanish maid employed in the 16th *arrondissement* of Paris in the 1960s now has her counterpart in the Filipina working in central Rome, or in the Dominican or Moroccan woman of the residential areas of Madrid (Colectivo Ioé 1991): the Bangladeshi street trader in the East End of London is now duplicated by the Moroccan carpet seller of central Barcelona or Lisbon. And although political discourse understandably highlights the role of clandestine migration in southern Europe we should remember that clandestinity has been a major issue elsewhere, for example in France (Marie 1983).

Certainly contemporary migration into southern Europe, in a situation of post-industrial economic change, has led to new forms of employment not found to the north, but such employment is in many respects even more marginalized than ever. The differences between southern European immigration today and migration into northern Europe in the past, although certainly present, are not overwhelming. However, it is not implausible to envisage a parallel political development towards highly controlled immigration regimes, such a development stemming from several pressures – public opinion, local political expediency, and the extra factor of pressure from other member states of the European Union to harmonize policies over an extremely sensitive issue.

As a concluding point, however, we should note that the significance of new migration into southern Europe is recognized only differentially as a "problem". The discussion in this chapter has highlighted the issues that have been discussed in academic research and which dominate current media and political discourse. Such research and discourse "problematizes" certain types of migration and ignores others. The concerns in Portugal, Italy, Spain and Greece are about migrants from less developed countries – seen as "outsiders" in one way or another. There is far less concern about the growing numbers of migrants from other advanced societies: these high-skilled, often circulatory, sometimes entrepreneurial, sometimes retirement migrants are not currently perceived as "problems" in any major way. Yet as was seen in the case of Spain (Table 10.6) there are not inconsiderable numbers of such movers, although survey evidence suggests that the public scarcely perceives this to be the case (Aguilera-Arilla et al. 1992).

Paradoxically, perhaps the most certain thing that can be said about future migration affecting southern Europe is that the numbers of foreign residents from these origins will continue to increase into the twenty-first century. Even if the flow from less developed countries were to be cut to a trickle, migration between the member states of the European Union is likely to increase. Is it conceivable that in twenty years' time it will be the British, Dutch and German immigrants who will be perceived as a problem in certain regions of southern Europe (such as the Algarve, the Mediterranean coast of Spain, or the Italian lakes) where they will have become culturally dominant through large-scale in-movement (and where their consumption demands might be met by Third World immigrants holding menial positions in the leisure industry and in trading)? Thirty years ago the German geographer Ritter (1966) described the transformation of the historic German magnet of settlement in the east (the *Drang nach Osten*) into the new southern magnet for tourism and recreation (the *Drang nach Süden*). Not only are the countries of the northern shore of the Mediterranean the first landfall for those from much of the world "South" arriving in Europe, they also constitute the European Sunbelt. These dual aspects of relative location will be of great significance in the development of migration over both the short and the medium term.

References

Aguilera-Arilla, M. J., M. P. Gonzalez Yanci, V. Rodríguez-Rodríguez 1992. Attitudes of Spanish population about foreign immigrants. Paper delivered at the conference on Mass Migration in Europe, Vienna, March.

Ahlburg, D. A. & C. J. De Vita 1992. New realities of the American family. *Population Bulletin* **47**(2), 1–44.

Amman, A. 1985. *L'evolution de la structure par âge de la population et politiques futures.* Strasbourg: Council of Europe.

Anderson, M. 1983. What is new about the modern family: an historical perspective. In *The family.* Occasional Paper 31 (1-16), OPCS, London.

Anon 1993. Labour force projections. *Employment Gazette* **101**, 139–48.

Arena, G. 1982. Lavoratori stranieri in Italia e a Roma. *Societá Geografica Italiana, Bollettino* **11**, 57–93.

Ballesteros, A. G. 1993. Unemployment: regional variations in age- and sex-specific rates. In *The changing population of Europe*, D. Noin & R. Woods (eds), 151–60. Oxford: Basil Blackwell.

Barnett, A. & P. Blaikie 1992. *AIDS in Africa: its present and future impact.* London: Pinter (Belhaven).

Barsotti, O. & L. Lecchini 1989. L'immigration des pays du Tiers-Monde en Italie. *Revue Européenne des Migrations Internationales* **5**, 45–63.

Beaverstock, J. 1990. New international labour markets. *Area* **22**, 151–8.

— 1991. Skilled international labour migration: an analysis of the geography of international secondments within large accountancy firms. *Environment and Planning A* **23**, 1133–46.

Berry, B. J. L. 1976. The counterurbanization process: urban America since 1970. In *Urbanization and counterurbanization*, B. J. L. Berry (ed.), 17–30. Beverly Hills, California: Sage.

— 1988. Migration reversals in perspective: the long-wave evidence. *International Regional Science Review* **11**, 245–52.

— 1991. *Long-wave rhythms in economic development and political behaviour.* Baltimore: Johns Hopkins University Press.

Bichot, J. 1993. *Quelles retraites en l'an 2000? Constat, analyses, solutions.* Paris: Armand Colin.

Birindelli, A. M. 1990. Gli aspetti quantitativi del fenomeno. In *La presenza straniera in Italia*, F. Labos (ed.), 50–62. Rome: Edizioni TER.

Böhning, W. R. 1972. *The migration of workers in the United Kingdom and the European Community.* Oxford: Oxford University Press.

Bone, M. 1986. Trends in single women's sexual behaviour in Scotland. *Population Trends* **43**, 7–14.

Bonifazi, C. 1993. From the Third World to Italy: the experiences of a new immigration country, between growth of push factors and containment policies. Paper presented at the General Conference of the International Union for the Scientific Study of Population, Montreal.

Bonifazi, C. & G. Gesano 1994. Immigrazione straniera tra regolazione dei flussi e politiche di accoglimento. In *Tendenze demografiche e politiche della popolazione Italiana*, A. Golini (ed.), 259–92. Bologna: Il Mulino.

Bourgeois-Pichat, J. 1965. Les facteurs de la fecondité dirigée. *Population* **3**, 383–424.

—1987. The unprecedented shortage of births in Europe. In *Below replacement fertility in industrial societies*, K. Davis, M. S. Bernstam & R. Ricardo Campbell (eds), 3–25. New York: The Population Council.

—1989. From the 20th to the 21st century: Europe and its population after the year 2000. *Population* (English selection I) **44**, 57–89.

Bouvier-Colle, M. H. & E. Jougla 1989. Etude européenne de la répartition géographique des causes de décès "évitables". In *Géographie et socioéconomie de la santé*, tome 2. Paris: CREDES.

Boyle, M., A. Findlay, E. Lelièvre & R. Paddison 1994. French investment and skill transfer in the United Kingdom. In *Population migration and the changing world order*, W. Gould & A. Findlay (eds), 47–65. Chichester: John Wiley.

Broard, N. & A. Lopez 1985. Cause of death in low mortality countries: a classification analysis. In *International Population Conference* (New Delhi 1985), vol. 2, 385–406 Liège: IUSSP.

Brochmann, G. 1993. Migration policies of destination countries. In *Political and demographic aspects of migration flows to Europe* [Population Studies 25], 105–28. Strasbourg: Council of Europe.

Campani, G. 1989. Du Tiers-Monde à l'Italie: une nouvelle immigration féminine. *Revue Européenne des Migrations Internationales* **5**, 29–49.

— 1993. Immigration and racism in southern Europe: the Italian case. *Ethnic and Racial Studies* **16**, 507–35.

Casas Torres, J. M. 1982. *Población, desarrollo y calidad de vida*. Madrid: Rialp.

Caselli, G. & V. Egidi 1981. *New trends in European mortality* [Population Studies 5]. Strasbourg: Council of Europe.

Casper, W. & S. Hermann 1991. The development of life expectancy in European countries. In *Socioeconomic differential mortality in industrialized societies*, vol. 7, 215–26. Paris: CICRED.

Castles, S. & M. J. Miller 1993. *The age of migration: international population movements in the modern world*. London: Macmillan.

Cavaco, C. 1993. A place in the sun: return migration and rural change in Portugal. In *Mass migration in Europe*, R. King (ed.), 174–94. London: Pinter (Belhaven).

CEC 1989, 1990, 1991, 1992, 1993. *Employment in Europe*. Luxembourg: Commission of the European Communities.

— 1991a. *Europe 2000: outlook for the development of the Community's territory*. Brussels: Commission of the European Communities.

— 1991b. *The regions in the 1990s: fourth periodic report on the social and economic situation and development of the regions of the Community*. Brussels/Luxembourg: Commission of the European Communities.

Cecchini, P. 1988. *The European challenge 1992: The benefits of a single market*. Aldershot: Gower.

Champion, A. G. (ed.) 1989. *Counterurbanization: The changing pace and nature of population deconcentration*. London: Edward Arnold.

— 1991. Changes in the spatial distribution of the European population. In *Seminar on Present Demographic Trends and Life Styles in Europe: proceedings*, 355–88. Strasbourg: Council of Europe.

— 1992. Urban and regional demographic trends in the developed world. *Urban Studies* **29**, 461–82.

— 1993. Geographical distribution and urbanization. In *The changing population of Europe*, D. Noin & R. Woods (eds), 23–37. Oxford: Basil Blackwell.

Champion, A. G. & P. D. Congdon 1992. Migration trends for the South: the emergence of a greater South East? In *Migration processes and patterns, volume 2: population redistribution in the United Kingdom*, J. Stillwell, P. Rees, P. Boden (eds), 178–203. London: Pinter (Belhaven).

Champion, A. G. & S. Illeris 1990. Population redistribution trends in western Europe. In *Unfamiliar territory: the reshaping of European geography*, M. Hebbert & J. C. Hansen (eds), 236–53. Aldershot: Avebury.

Chaunu, P. 1974. *Histoire, science sociale. La durée, l'espace et l'homme à l'époque moderne*. Paris: SEDES.

Cheshire, P. C. & D. G. Hay 1989. *Urban problems in western Europe*. London: Unwin Hyman.

Chesnais, J.C. 1991a. *La population du monde de l'Antiquité à 2050*. Paris: Bordas.

—1991b. The USSR emigration: past, present and future. Paper presented to the International Conference on Migration, Rome, March 1991.

Clarke, J. I. 1972. *Population geography*. Oxford: Pergamon Press.

Clarke, J. I. & W. B. Fisher (eds) 1972. *Populations of the Middle East and North Africa: a geographical approach*. London: London University Press.

Cliff, A. D. & P. Haggett 1988. *Atlas of disease distributions*. Oxford: Basil Blackwell.

Cliquet, R. (ed.) 1993. *The future of Europe's population: a scenario approach*. Strasbourg: Council of Europe Press.

Cochrane, S. G. & D. R. Vining 1988. Recent trends in migration between core and peripheral regions in developed and advanced developing countries. *International Regional Science Review* **11**, 215–44.

Cole, J. P. & I. V. Filatotchev 1992. Some observations on migration within and from the former USSR in the 1990s. *Post-Soviet Geography* **7**, 432–53.

Colectivo Ioé 1991. *Trabajadoras extranjeras de servicio doméstico en Madrid*. Working Paper 51S, International Labour Organization, Geneva.

— 1993. Hacia un análisis sociológico de la inmigración. Extranjeros en la comunidad de Madrid. *Política y Sociedad* **12**, 61–78.

Coleman, D. & J. Salt 1993. *The British population*. Oxford: Oxford University Press.

Collicelli, C. & S. di Cori 1986. L'immigrazione straniera in Italia nel contesto delle problematiche migratorie internazionali. *Studi Emigrazione* **82–3**, 429–36.

Compton, P. 1976. Migration in eastern Europe. In *Migration in post-war Europe: geographical essays*, J. Salt & H. Clout (eds), 168–215. Oxford: Oxford University Press.

Comunicación del Gobierno al Congreso de los Diputados 1991. Situación de los extranjeros en España. Líneas básicas de la política española de extranjería. *Revista de Economía y Sociología del Trabajo* **11**, 263–80.

Conroy Jackson, P. 1991. *The context of women's initiatives on local employment*. Local Government Centre, University of Warwick.

Cooke, P. (ed.) 1989. *Localities: the changing urban face of Britain*. London: Unwin Hyman.

Coombes, M., R. Dalla Longa, S. Raybould 1989. Counterurbanization in Britain and Italy: A comparative critique of the concept, causation and evidence. *Progress in Planning* **32**(1), 1–70.

Council of Europe, 1993. *Recent demographic developments in Europe and North America 1992*. Strasbourg: Council of Europe Press.

— 1994. *Recent demographic developments in Europe 1993*. Strasbourg: Council of Europe Press.

Cross, D. F. W. 1990. *Counterurbanization in England and Wales*. Aldershot: Avebury.

Cruijsen, H. 1991. *Fertility in the European Community*. In *Human resources at the dawn of the 21st century, Conference proceedings*. Luxembourg: Eurostat.

Davidoff, L. & C. Hall 1987. *Family fortunes*. London: Hutchinson.

Decroly, J-M. & J-P. Grimmeau 1991. Variations intercommunales de la mortalité par âge en Belgique. *Espace, Populations, Sociétés*, 1991/1, 75–83.

Decroly, J-M. & J. Vanlaer 1991. *Atlas de la population Européenne*. Brussels: Editions de l'Université de Bruxelles.

dell'Agnese, E. 1993. Gli albanesi in Italia: un caso di immigrazione "controllata". Paper presented at the British–Italian Seminar on Population Geography, Cagliari (Italy), September 1993.

Del Campo, S. 1990. Current family policy in Spain. In *Family policies in EEC countries*, W. Dumon (ed.), 335–49. Luxembourg: Office for Official Publications of the European Community.

Dematteis, G. & P. Petsimeris 1989. Italy: counterurbanization as a transitional phase in settlement reorganization. In *Counterurbanization*, A. G. Champion (ed.), 187–206. London: Edward Arnold.

Department of Employment 1991. Labour force trends: the next decade. *Employment Gazette* **99**, 269–80.

Deroure, F. 1992. *Professional mobility in Europe*. Brussels: EC Directorate General for Employment, Industrial Relations and Social Affairs.

Desplanques, G. 1984. L'inégalité sociale devant la mort. *Economie et Statistique* **179**, 26–49.

Dewdney, J. & P. White 1986. Portugal. In *West European population change*, A. Findlay & P. White (eds), 187–207. Beckenham, England: Croom Helm.

Dicken, P. 1992. *Global shift*. New York: Harper & Row.

Dirección General de Migraciones 1993. *Anuario de migraciones*. Madrid: Ministerio de Trabajo y Seguridad Social.

Donaldson, L. 1991. *Fertility transition. The social dynamics of population change*. Cambridge, Massachusetts: Basil Blackwell.

Dormor, D. J. 1992. *The relationship revolution*. London: One Plus One.

Dunlop, J. B. 1993. Will a large-scale migration of Russians to the Russian Republic take place over the current decade? *International Migration Review* **27**, 605–29.

Dunnell, K. 1979. *Family formation 1976*. London: OPCS.

Dutourd, J. 1975. *2024*. Paris: Gallimard.

Easterbrook, P. & D. FitzSimons 1992. Is the collapse of communism fuelling HIV? *New Scientist* (22 August), 11–12.

Eaton, M. 1993. Foreign residents and illegal immigrants: os negros em Portugal. *Ethnic and Racial Studies* **16**, 536–62.

Ermisch, J. 1990. *Fewer babies, longer lives*. York: Joseph Rowntree Foundation.

Estébanez Alvarez, J. & R. Puyol Antolín 1973. Los movimentos migratorios españoles durante el decenio, 1961–1970. *Geographica* **15**, 105–42.

European Foundation for the Improvement of Living and Working Conditions 1990. *Mobility and social cohesion in the European Community*. Luxembourg: Commission of the European Communities.

Eurostat 1988. *Demographic and labour force analysis based on Eurostat data banks.* Luxembourg: Commission of the European Communities.

—1991a. *Demographic indicators of the Community.* Luxembourg: Commission of the European Communities.

— 1991b. *Eurostat rapid reports.* Luxembourg: Commission of the European Communities.

—1991c. *Demographic statistics 1991.* Luxembourg: Statistical Office of the European Communities.

—1992. *Demographic statistics 1992.* Luxembourg: Statistical Office of the European Communities.

Faus-Pujol, M. C. 1991. Differential fertility in Spain, In *The geographical approach to fertility,* J. Bähr & P. Gans (eds), 129–49. Kiel: Kieler Geographische Schriften.

Fielding, A. J. 1982. Counterurbanization in western Europe. *Progress in Planning* 17, 1–52.

— 1986. Counterurbanization in western Europe. In *West European population change,* A. Findlay & P. White (eds), 35–49. London: Croom Helm.

— 1990. Counterurbanization: threat or blessing? In *Western Europe: challenge and change,* D. Pinder (ed.), 226–39. London: Pinter (Belhaven).

— 1993a. Migration and the metropolis: an empirical and theoretical analysis of interregional migration to and from South East England. *Progress in Planning* 39, 73–166.

— 1993b. Mass migration and economic restructuring. In *Mass migration in Europe,* R. King (ed.), 7–18. London: Pinter (Belhaven).

Findlay, A. 1988. From settlers to skilled transients. *Geoforum* 19, 401–10.

— 1990. A migration channels approach to the study of high-level manpower movements: a theoretical perspective. *International Migration* 28, 15–24.

— 1992. *The economic impact of immigration to the United Kingdom.* Applied Population Research Unit Discussion Paper 92/5, University of Glasgow.

— 1993. New technology, high-level labour movements and the concept of the brain drain. In *The changing course of international migration,* 149–59. Paris: OECD.

Findlay, A. & W. Gould 1989. Skilled international migration: a research agenda. *Area* 21, 3–11.

Findlay, A. & R. Rogerson 1993. Migration, places and quality of life: voting with their feet? In *Population matters: the local dimension,* A. G. Champion (ed.), 33–50. London: Paul Chapman.

Fondazione G. Brodolini 1992. *Rapporto sulla cooperazione e le politiche migratorie.* Rome: Cooperazione Italia.

Ford, R. 1992. *Migration and stress among corporate employees.* PhD thesis, University College London, University of London.

Fox, J. 1989. *Health inequalities in European countries.* Aldershot: Gower.

Frey, W. H. 1987. Migration and depopulation of the metropolis: regional restructuring or rural renaissance? *American Sociological Review* 52, 240–57.

— 1993. The new urban revival in the USA, *Urban Studies* 30, 741–74.

Frey, W. H. & A. Speare 1992. The revival of metropolitan population growth in the United States: an assessment of findings from the 1990 Census. *Population and Development Review* 18, 129–46.

Fröbel, F., J. Heinrichs, O. Kreye 1980. *The new international division of labour.* Cambridge: Cambridge University Press.

Fukuyama, F. 1989. The end of history? *The National Interest* 16, 3–18.

— 1992. *The end of history and the last man.* London: Hamish Hamilton.

193

Gesano, G. & F. Heins (1994). Trasformazione demografica e interventi sul mercato. In *Tendenze demografiche e politiche della popolazione Italiana*, A. Golini (ed.), 227-58. Bologna: Il Mulino.

Geyer, H. S. & T. Kontuly 1993. A theoretical foundation for the concept of differential urbanization. *International Regional Science Review* 15, 157-77.

Glezer, O. B. & V. N. Streletsky 1991. Reclamaciones territoriales y conflictos étnicos en el proceso de desintegración de la Unión Soviética. *Estudios Geograficos* 52, 421-38.

Goldin, I., O. Knudsen, D. van der Mensbrugghe 1993. *Trade liberalisation: global economic implications*. Paris: OECD.

Golini, A. 1987. Famille et ménage dans l'Italie récente. *Population* 42, 699-714.

—1994. Le tendenze demografiche dell'Italia in un quadro europeo – analisi e problemi per una politica per la popolazione. In *Tendenze demografiche e politiche della popolazione Italiana*, A. Golini (ed.), 17-78. Bologna: Il Mulino.

— C. Bonifazi & A. Righi 1993. A general framework for the European migration system in the 1990s. In *The new geography of European migrations*, R. King (ed.), 67-83. London: Pinter (Belhaven).

Gorer, G. 1971. *Sex and marriage in England today*. London: Thomas Nelson.

Gould, P. 1993. *The slow plague: a geography of the AIDS pandemic*. Oxford: Basil Blackwell.

Gozálvez, V. 1990. El reciente incremento de la población extranjera en España y su incidencia laboral. *Investigaciones Geográficas* 8, 7-36.

Grass, G. 1980. *Kopfgeburten*. Darmstadt: Luchterhand. [Published in English translation (1984)as *Headbirths, or the Germans are dying out*. Harmondsworth: Penguin.]

Gray, J. 1993. *Beyond the New Right: markets, government and the common environment*. London: Routledge.

Grebenik, E. 1991. Demographic research in Britain 1936–1986. In *Population research in Britain*, M. Murphy & J. Hobcraft (eds), 3–30. London: Population Investigation Committee.

Green, A. E. 1992. Spatial aspects of the SEM scenarios. In *Women's employment: Britain in the Single European Market*, R. Lindley (ed.), 33–55. London: HMSO.

— 1993. *Changing female economic activity rates: issues and implications*. Institute for Employment Research, University of Warwick.

Green, A. & D. Owen 1991. Local labour supply and demand interactions in Britain during the 1980s. *Regional Studies* 25, 295–314.

—1993. Fall-out from the demographic time-bomb: a spatial perspective on the labour force effects of the "baby-bust". In *Population matters: the local dimension*, A. G. Champion (ed.), 83–100. London: Paul Chapman.

Gregory, D. D. 1978. *La Odisea Andaluza: una emigración hacia Europa*. Madrid: Tecnos.

Gritsai, O. & A. Treivish 1990. Stadial concept of regional development: centre and periphery in Europe. *Geographische Zeitschrift* 78, 65–77.

Guha, A. 78. Intra-COMECON manpower migration. *International Migration* 16, 52–65.

Guillon, M. 1989. Réfugiés et immigrés de l'Europe de l'Est. *Revue Européenne des Migrations Internationales* 5, 133–8.

Guilmot, P. 1978. The demographic background in *Population decline in Europe*, Council of Europe, 3-13. London: Edward Arnold.

Habakkuk, H. J. 1971. *Population growth and economic development since 1750*. Leicester: Leicester University Press.

Hajnal, J. 1965. European marriage patterns in perspective. In *Population in history*, D. V. Glass

& D. E. C. Eversley (eds), 101–143. London: Edward Arnold.

Hakim, C. 1993. The myth of rising female employment. *Work, Employment and Society* **7**, 97–120.

Hall, P. 1971. Spatial structure of metropolitan England and Wales. In *Spatial policy problems of the British economy*, M. Chisholm & G. Manners (eds), 95–125. Cambridge: Cambridge University Press.

— 1987. The geography of high technology: an Anglo-American comparison. In *The spatial impact of technological change*, J. Brotchie, P. Hall, P. W. Newton (eds), 141–56. London: Croom Helm.

Hall, P. & D. Hay 1980. *Growth centres in the European urban system*. London: Heinemann.

Hall, R. 1988. Recent patterns and trends in western European households at national and regional scales. *Espace Populations Sociétés* **1**, 13–32.

— 1993. Family structures. In *The changing Population of Europe*, D. Noin & R. Woods (eds), 100–126. Oxford: Basil Blackwell.

Hamnett, C. 1994. Social polarization in global cities: theory and evidence. *Urban Studies* **31**, 401–24.

Haughton, G. 1990. Skills shortage and the demographic time-bomb: labour market segmentation and the geography of labour. *Area* **22**, 339–45.

Heath, S. & A. Dale 1994. Household and family formation in Great Britain: the ethnic dimension. *Population Trends*, **77**, 5–13.

Heitman, S. 1987. *The third Soviet emigration: Jewish, German and Armenian emigration from the USSR since World War II*. Berichte des Bundesinstituts für Ostwissenschaftliche und Internationale Studien, Cologne.

Herdegen, G. 1989. Aussiedler in der Bundesrepublik Deutschland: Einstellen und aktuelle Ansichten der Bundesbürger. *Informationen zur Raumentwicklung* **5**, 331–41.

Higueras Arnal, A. 1991. Fertility and social change in Spain (1975–1987). In *The geographical approach to fertility*, J. Bähr & P.Gans (eds), 121–8. Kiel: Kieler Geographische Schriften.

Hills, J. 1993. *The future of welfare: a guide to the debate*. York: Joseph Rowntree Foundation.

Hockenos, P. 1994. *Free to hate: the rise of the Right in post-communist eastern Europe*. London: Routledge.

Hoffmann-Nowotny H-J. & B. Fux 1991. Present demographic trends in Europe. In *Seminar on present demographic trends and lifestyles in Europe*, 31–97. Strasbourg: Council of Europe.

Holland, W. W. (ed.) 1988. *European Community atlas of "avoidable death"*. Oxford: Oxford University Press.

Hopflinger, F. 1985. Changing marriage behaviour: some European comparisons. *Genus* **41**, 41–64.

— 1991. The future of household and family structures in Europe. In *Seminar on present demographic trends and life styles in Europe: proceedings*, 289-338. Strasbourg: Council of Europe.

House, J. W. 1982. *Frontier on the Rio Grande: a political geography of development and social deprivation*. Oxford: Oxford University Press.

Hovanessian, M. 1988. Soixante ans de présence arménienne en région Parisienne (le cas d'Issy-les-Moulineaux). *Revue Européenne des Migrations Internationales* **4**, 73–95.

Illeris, S. 1988. Counterurbanization revisited: the new map of population distribution in central and northwestern Europe. In *Urbanization and urban development*, M. Bannon, L. Bourne, R. Sinclair (eds), 1-16. Dublin: University College.

— 1992. Urban and regional development in western Europe in the 1990s: a mosaic rather than the triumph of the "Blue Banana". *Scandinavian Housing & Planning Research* **9**, 201–15.

Independent 1993. Parents under pressure, 5 (28 July).

IER 1991. *Review of the economy and employment*. Coventry: Institute for Employment Research, University of Warwick.

— 1993. *Review of the economy and employment*. Coventry: Institute for Employment Research, University of Warwick.

Izquierdo, A. 1992a. Las encuestas contra la inmigración. Paper delivered at the conference entitled Jornadas Sobre Racismo, Xenofobia y Diversidad Cultura, organized by Facultad de Filosofía y Letras de la Universidad Autónoma de Madrid y la Facultad de Ciencias Políticas y Sociología del Universidad Complutense de Madrid, March 1992.

— 1992b. *La inmigración en España 1980–1990*. Madrid: Ministerio de Trabajo y Seguridad Social.

James, P. D. 1993. *The children of men*. Harmondsworth: Penguin.

Johnston, R. J. 1994. One world, millions of places: the end of history and the ascendancy of geography. *Political Geography* **13**, 111–21.

Jones, H. 1991. The French Census 1990: the southward drift continues. *Geography* **76**, 358–61.

Jones, P. N. & M. T. Wild 1992. Western Germany's "Third Wave" of migrants: the arrival of the Aussiedler. *Geoforum* **23**, 1–12.

Joshi, H. 1989. The changing form of women's economic dependency. In *The changing population of Britain*, H. Joshi (ed.), 157–76. Oxford: Basil Blackwell.

de Jouvenal, H. 1989. *Europe's ageing population: trends and challenges to 2025*. Guildford: Futures.

Kane, T. T. 1986. The fertility and assimilation of guest-worker populations in the Federal Republic of Germany, 1961–81. *Zeitschrift für Bevölkerungswissenschaft* **12**, 99–131.

Karger, A. 1988. Ethnischer Wandel in Lettland. *Geographische Rundschau* **40**, 34–7.

Keeble, D. 1989. Core–periphery disparities, recession and new regional dynamisms in the European Community. *Geography* **74**, 1–11.

Kennedy, P. 1988. *The rise and fall of the Great Powers: economic change and military conflict from 1500 to 2000*. London: Unwin Hyman.

Kennedy, P. 1993. *Preparing for the twenty-first century*. London: HarperCollins.

Keyfitz, N. 1993. Culture and the birth rate. In *Conference proceedings Vol. 1, IUSSP International Population Conference, Montreal*.

Kiernan, K. & V. Estaugh 1993. *Cohabitation*. London: Family Policy Studies Centre.

King, R. L. 1976. The evolution of international labour migration movements concerning the EEC. *Tijdschrift voor Economische en Sociale Geografie* **67**, 66–82.

— (ed.) 1986. *Return migration and regional economic problems*. Beckenham: Croom Helm.

— 1991. Europe's metamorphic demographic map. *Town and Country Planning* **60**, 111–13.

— 1993a. Recent immigration to Italy: character, causes and consequences. *Geojournal* **30**, 283–92.

— 1993b. *Mass migration in Europe: the legacy and the future*. London: Pinter (Belhaven).

— 1993c. European international migration 1945–90. In *Mass migration in Europe*, R. King (ed.), 19–39. London: Pinter (Belhaven).

King, R. & K. Rybaczuk 1993. Southern Europe and the international division of labour:

from emigration to immigration. In *The new geography of international migrations*, R. King (ed.) 175–206. London: Pinter (Belhaven).

Klaassen, L. H., W. T. M. Molle, J. H. P. Paelinck (eds) 1981. *Dynamics of urban development*. Aldershot: Gower.

Korcelli, P. 1988. Migration trends and regional labour market change in Poland. *Geographia Polonica* **54**, 5–17.

— 1992. International migrations in Europe: Polish perspectives for the 1990s. *International Migration Review* **26**, 292–304.

Kunst, A. et al. 1988. Medical care and regional mortality differences within the countries of the European Community. *European Journal of Population* **4**, 223–45.

Kunzmann, K. R. & M. Wegener 1991. *The pattern of urbanization in western Europe 1960– 1990*. Berichte aus dem Institut für Raumplanung 28, Institut für Raumplanung, Universität Dortmund.

Laqueur, W. 1994. *Black Hundred: the rise of the extreme Right in Russia*. New York: Harper-Collins.

Laslett, P. 1980. Introduction: comparing illegitimacy over time and between cultures. In *Bastardy and its comparative history*, P. Laslett, K. Oosterveen, R. M. Smith (eds), 1–70. London: Edward Arnold.

Leitner, H. 1990. Informal work on the streets of Vienna: the foreign newspaper vendors. *Espace, Populations, Sociétés* (2), 221–9.

Leloup, Y. 1972. L'émigration portugaise dans le monde et ses conséquences pour le Portugal. *Revue de Géographie de Lyon* **47**, 59–76.

Leridon, H. 1990. Cohabitation, marriage, separation: an analysis of life histories of French cohorts from 1968 to 1985. *Population Studies* **44**, 127–44.

Leridon, H. & C. Villeneuve-Gokalp 1988. Les nouveaux couples: nombre, caractéristiques et attitudes. *Population* **43**, 331–74.

Lesthaeghe, R. & J. Surkyn 1988. Cultural dynamics and economic theories of fertility change. *Population and Development Review* **14**, 1–45.

Levy, M. L. 1989. L'enfant européen. *Population et sociétés* **234**, 1–4.

Lianos, T. P. 1975. Flows of Greek out-migration and return migration. *International Migration* **13**, 119–33.

Lichtenberger, E. 1984. *Gastarbeiter – Leben in zwei Gesellschaften*. Vienna: Hermann Böhlau.

Lindley, R. M. 1990. *Demographic change and labour market policies*. Institute for Employment Research, University of Warwick.

— R. A. Wilson, E. Villagomez 1991. *Labour market prospects for the third age*. Institute for Employment Research, University of Warwick.

López, B. 1993. *La inmigración Magrebí en España: el retorno de los Moriscos*. Madrid: Mapfre.

Löytönen, M. 1991. The spatial diffusion of human immunodeficiency virus type 1 in Finland, 1982–1997. *Association of American Geographers, Annals* **81**, 127–51.

— 1992. HIV and Russia – are they sitting on a bomb? In *Program and Abstracts of the International Medical Geography Symposium*, Charlotte, North Carolina, 4–7 August, 22.

Luker, K. 1984. *Abortion and politics of motherhood*. Berkeley: University of California Press.

Lutz, W. (ed.) 1991. *Future demographic trends in Europe and North America*. London: Academic Press.

— (ed.) 1994. *The future of world population: what can we assume today?* London: Earthscan.

Lutz, W. & C. Prinz 1992. What difference do alternative immigration and integration levels make to western Europe? *European Journal of Population* **8**, 341–61.

— 1993. Modelling future immigration and integration in western Europe. In *The new geography of European migrations*, R. King (ed.), 83–99. London: Pinter (Belhaven).

Maalouf, A. 1992. *Le premier siècle après Béatrice*. Paris: Grasset et Fasquelle.

Mann, J. 1991. AIDS – the second decade: a global perspective. *Journal of Infectious Diseases* **165**, 245–50.

Marie, C. 1983. L'immigration clandestine en France. *Hommes et Migrations* **1059**, 4–21.

Martin, P. L. 1991. *The unfinished story: Turkish labour migration to western Europe, with special reference to the Federal Republic of Germany*. Geneva: International Labour Office.

Mason, T. 1978. Residential succession, community facilities and urban renewal in Cheetham Hill, Manchester. *New Community* **6**, 78–87.

Masser, I., O. Svidén, M. Wegener 1992. *The geography of Europe's futures*. London: Pinter (Belhaven).

Matthiessen, P. C. 1991 Conclusions by the general rapporteur. In *Seminar on present demographic trends and lifestyles in Europe*, 413–21. Strasbourg: Council of Europe.

McLaren, M. 1990. *A history of contraception*. Oxford: Basil Blackwell.

McPherson, A. 1993. *Scottish skills in the Global Economy*. Applied Population Research Unit Discussion Paper 93/2, University of Glasgow.

McRae, S. 1993. *Cohabiting mothers*. London: Policy Studies Institute.

Meulders, D., C. Hecq & R. Plasman 1992. An assessment of the European evidence on the employment of women and 1992. In *Women's employment: Britain in the Single European Market*, R. Lindley (ed.), 158–80. London: HMSO.

Miller, M. W. 1994. Moldova: a new nation-state. Paper presented to the Institute of British Geographers Annual Conference, Nottingham.

Mingione, E. 1985. Marginale e povero: il nuovo immigrato in Italia. *Politica ed Economia* **6**, 61–4.

Monnier, A. 1990. The demographic situation of Europe and the developed countries overseas: an annual report. *Population* (English selection) **2**, 234–42.

Montanari, A. & A. Cortese 1993. Third World immigrants in Italy. In *Mass migrations in Europe: the legacy and the future*, R. King (ed.), 275–92. London: Pinter (Belhaven).

Morokvasic, M. 1992. La guerre et les réfugiés dans l'ex-Yougoslavie. *Revue Européenne des Migrations Internationales* **8**, 5–25.

Morsa, J. 1979. *Socio-economic factors affecting fertility and motivation for parenthood*. Population Studies 3, Strasbourg, Council of Europe.

Muñoz-Perez, F. 1987. Le déclin de la fécondité dans le Sud de l'Europe. *Population* **6**, 911–41.

Muñoz-Perez, F. & A. Izquierdo Escribano 1989. L'Espagne: pays d'immigration. *Population* **44**, 257–89.

Murphy, M. & A. Berrington 1993. Household change in the 1980s: a review. *Population Trends* **73**, 18–26.

National Children's Bureau 1993. *Life at thirty-three*. London: National Children's Bureau.

National Economic Development Office/Training Commission 1988. *Young people and the labour market: a challenge for the 1990s*. London: NEDO.

National Economic Development Office/Training Agency 1989. *Defusing the demographic time bomb*. London: NEDO.

Naumkin, V. V. (ed.) 1994. *Central Asia and Transcaucasia: ethnicity and conflict*. London: Greenwood Press.

Noin, D. 1983. *La transition démographique dans le monde*. Paris: Presses Universitaries de France.

— 1991. *Atlas de la population mondiale*. Paris: Reclus-La Documentation Française.

Öberg, S. & H. Boubnova 1993. Ethnicity, nationality and migration potentials in eastern Europe. In *Mass migration in Europe: the legacy and the future*, R. King (ed.), 234–56. London: Pinter (Belhaven).

OECD 1990. *Employment Outlook 1990*. Paris: OECD.

— 1992. *Trends in international migration*. Paris: OECD.

OPCS, 1986. Births by birthplace of parents, 1985. *OPCS Monitor* FM1 86/5.

Owen, D. W. 1994. *Black people: social and economic characteristics*. Centre for Research in Ethnic Relations, University of Warwick.

Parkinson, M. 1991. European cities in the 1990s: problems and prospects. In *The future of cities in Britain and Canada*, I. Jackson (ed.), 67–87. Aldershot: Dartmouth.

Pattie, C. J. 1994. Forgetting Fukuyama: new spaces of politics. *Environment and Planning* A **26**, 1007–10.

Pope, S. & C. W. Mueller 1976. The intergenerational transmission of marital instability: comparisons by race and sex. *Journal of Social Issues* **32**, 49-66.

Poppel, F. van 1979. Regional differences in mortality in western and northern Europe: a review of the situation in the 1970s. *Proceedings of the Meeting on Socioeconomic Determinants and Consequences of Mortality* (Mexico 1979). New York/Geneva: United Nations/World Health Organisation.

Poulain, M. 1990. Une méthodologie pour faciliter la cartographie des niveaux de mortalité en l'absence de données sur les décès par âge. *Espace, Populations, Sociétés* 1990/3, 387–91.

Poulain, M., M. Debuisson, T. Eggericks 1991. *Proposals for the harmonization of European Community statistics on international migration*. Institute of Demography, Catholic University of Louvain.

Pumares, P. 1993. Factores de la estructura ocupacional de los inmigrantes extranjeros. Cuartas Jornadas de la Población Española, Tenerife, June 1993, 467–74.

Pumares, P. & A. Barroso 1993. *El Grupo de Discusión Aplicado al Análisis de las Actitudes de los Españoles Hacia la Inmigración (II): Análisis de los Grupos*. Documento de Trabajo N.15, Departamento de Estudios Urbanos y Territoriales, CSIC.

Ratledge, C., J. Stanford, J. M. Crange (eds) 1989. *The biology of the mycobacteria*, vol. 3: *Clinical aspects of mycobacterial disease*. London: Academic Press.

RECLUS 1989. *Les villes européennes: rapport pour la DATAR*. Montpellier: Maison de la Géographie.

Redei, M. 1992. Displaced persons in a new host country. Paper presented to the conference on Mass Migration in Europe, Vienna, March 1992.

Rees, P., J. Stillwell, A. Convey 1992. *Intra-Community migration and its impact on the demographic structure at the regional level*. Working Paper 92/1, Department of Geography, University of Leeds.

Rein, M. & K. Jacobs 1993. Ageing and employment trends: a comparative analysis for OECD countries. In *Labour markets in an ageing Europe*, P. Johnson & K. F. Zimmermann (eds), 53–76. Cambridge: Cambridge University Press.

Rhode, B. 1991. *East–West migration / Brain Drain*. Brussels: European Co-operation in the Field of Scientific and Technical Research (COST).

Ritter, W. 1966. *Fremdenverkehr in Europa*. Leiden: Elsevier.

Rodríguez, V. 1993. Problems in the definition of skilled migration and figures of skilled migration in Spain. Paper presented at the conference on Skilled and Highly Skilled Migration, Latina (Italy), October.

Rowland, R. H. 1988. Union republic migration trends in the USSR during the 1980s. *Soviet Geography* **29**, 809–29.

Rowley, G. 1992. Palestinian refugees: empirical and qualitative considerations. *GeoJournal* **27**, 217–27.

Sadler, D., A. Swain, R. Hudson 1993. The automobile industry and eastern Europe: new production strategies or old solutions? *Area* **25**, 339–49.

Sakkeus, L. 1992. Migration trends in Estonia: formation of the foreign-born population. Paper presented to the conference Mass Migration in Europe, Vienna, March 1992.

Salt, J. 1987. The SOPEMI experience. *International Migration Review* **21**, 1067–73.

— 1988. Highly skilled migrants, careers and international labour markets. *Geoforum* **19**, 387–99.

— 1993. *Migration and population change in Europe*. Research Papers 19, UNIDIR,Geneva.

Salt, J. & A. Findlay 1989. International migration of highly skilled manpower: theoretical and development issues. In *The impact of international migration on developing countries*, A. Appleyard (ed.), 159–81. Paris: OECD.

Salt, J. & R. Ford 1993. Skilled international migration in Europe. In *Mass migration in Europe*, R. King (ed.), 293–309. London: Pinter (Belhaven).

Satmarescu, G. D. 1975. The changing population structure of the population of Transylvania. *East European Quarterly* **25**, 425–47.

Shannon, G. W., G. Pyle, R. L. Bashshur 1991. *The geography of AIDS*. New York: Guildford.

Shorter, E. 1976. *The making of the modern family*. London: Collins.

Shuttleworth, I. 1993. Irish graduate emigration: the mobility of qualified manpower in the context of peripherality. In *Mass migration in Europe*, R. King (ed.), 310–26. London: Pinter (Belhaven).

Simon, G. 1993. Internal migration and mobility. In *The changing population of Europe*, D. Noin & R. Woods (eds), 170–84. Oxford: Basil Blackwell.

Singer, M. & A. Wildavsky 1993. *The real world order: zones of peace / zones of turmoil*. Chatham, New Jersey: Chatham House.

Smallman-Raynor M., A. Cliff, P. Haggett 1992. *Atlas of AIDS*. Oxford: Blackwell.

Smith, G. (ed.) 1990. *The nationalities question in the Soviet Union*. Harlow: Longman.

Sorel, A. 1974. *4° Mundo: emigración Española en Europa*. Madrid: Zero.

Starovojtova, G. V. 1987. *Etniceskaja Gruppa v Sovremennom Sovetskom Gorode*. Leningrad: Nauka.

Sundstom, M. & F. P. Stafford 1992. Female labour force participation, fertility and public policy in Sweden. *European Journal of Population* **8**, 199–215.

Sysdem 1990. *Sysdem Papers 4: Euro-labour markets: the prospects for integration*. Brussels: CEC.

Ter Minassian, A. 1989. La diaspora Arménienne. *Hérodote* **53**, 123–57.

Treibel, A. 1994. Le "sentiment du nous" en Allemagne. *Revue Européenne des Migrations Internationales* **10**, 57–72.

Trollope, A. 1990 (1882). *The fixed period* [republished]. Ann Arbor: University of Michigan Press.

UCC 1991. *Cadres en Europe*. Paris: CSA [Conseils, Sondages, Analyses].

Ulack, R. & W. F. Skinner (eds) 1991. *AIDS and the social sciences*. Lexington: University Press of Kentucky.

United Nations 1991. *World population prospects 1990*. New York: United Nations.

— 1992. *World urbanization prospects 1992*. New York: United Nations.

Valkonen, T. 1987. Social inequality in the face of death. European Population Conference, Plenaries, (Jyväskylä, Finland), 201–61.

Vallin, J. 1984. Politiques de santé et mortalité dans les pays industrialisés. *Espace, Populations, Sociétés* 1984/3, 13–31.

van de Kaa, D. J. 1987. Europe's second demographic transition. *Population Bulletin* **42**(1), 1–57.

van den Berg, L., R. Drewett, L. H. Klaassen, A. Rossi, C. H. T. Vijverberg 1982. *Urban Europe, volume 1: a study of growth and decline*. Oxford: Pergamon.

Vandermotten, C. 1993. The geography of employment. In *The changing population of Europe*, D. Noin & R. Woods (eds), 135–50. Oxford: Basil Blackwell.

Vining, D. R. & R. Pallone 1982. Migration between core and periphery regions: a description and tentative explanation of patterns of 22 countries. *Geoforum* **13**, 339–410.

Vishnevsky, A. & J. Zayonchkovskaya 1992. Emigration from the USSR: the fourth wave. Paper presented at the conference on Mass Migration in Europe, Vienna, March 1992.

Wallerstein, I. 1991. *Geopolitics and geoculture: essays on the changing world-system*. Cambridge: Cambridge University Press.

Wellings, K., J. Field, L. Whitaker 1994. Sexual attitudes. In *Sexual behaviour in Britain*, K. Wellings, J. Field, A. M. Johnson, J. Wadsworth (eds), 230–74. London: Penguin.

White, P. 1986. International migration in the 1970s: revolution or evolution? In *West European population change*, A. Findlay & P. White (eds), 50–80. Beckenham, England: Croom Helm.

White, P. 1993a. Ethnic minority communities in Europe. In *The changing population of Europe*, D. Noin & R. Woods (eds), 206–225. Oxford: Basil Blackwell.

— 1993b. The social geography of immigrants in European cities: the geography of arrival. In *The new geography of European migrations*, R. King (ed.), 47–66. London: Pinter (Belhaven).

White, P. (1994). Migration research. In *Population dynamics in Europe: current issues in population geography*, P. Hooimeijer et al. (eds), 53–68. Utrecht: Netherlands Geographical Studies.

Willekens, F. 1994. Monitoring international migration flows in Europe: towards a statistical database combining data from different sources. *European Journal of Population* **10**, 1–42.

Wils, A. B. 1991. Survey of immigration trends and assumptions about future migration. In *Future demographic trends in Europe and North America: what can we assume today?*, W. Lutz (ed.), 82–96. London: Academic Press.

World Bank 1992. *World development report 1992*. Oxford: Oxford University Press.

World Health Organisation 1993. The current global situation of the HIV/AIDS pandemic. *Weekly Epidemiological Record* (3), 11.

Wrigley, E. A. 1966. Family limitation in preindustrial England. *Economic History Review* (2nd series) **XIX**, 82–109

Zayonchkovskaya, J., A. Kocharyan, G. Vitkovskaya 1991. Forced migrations in the context of ethnic processes in the USSR. Paper presented at the conference on The Refugee Crisis, Kings College, London, September 1991.

Index